Moving and Growing

Becca Heddle

Series editor Sue Palmer

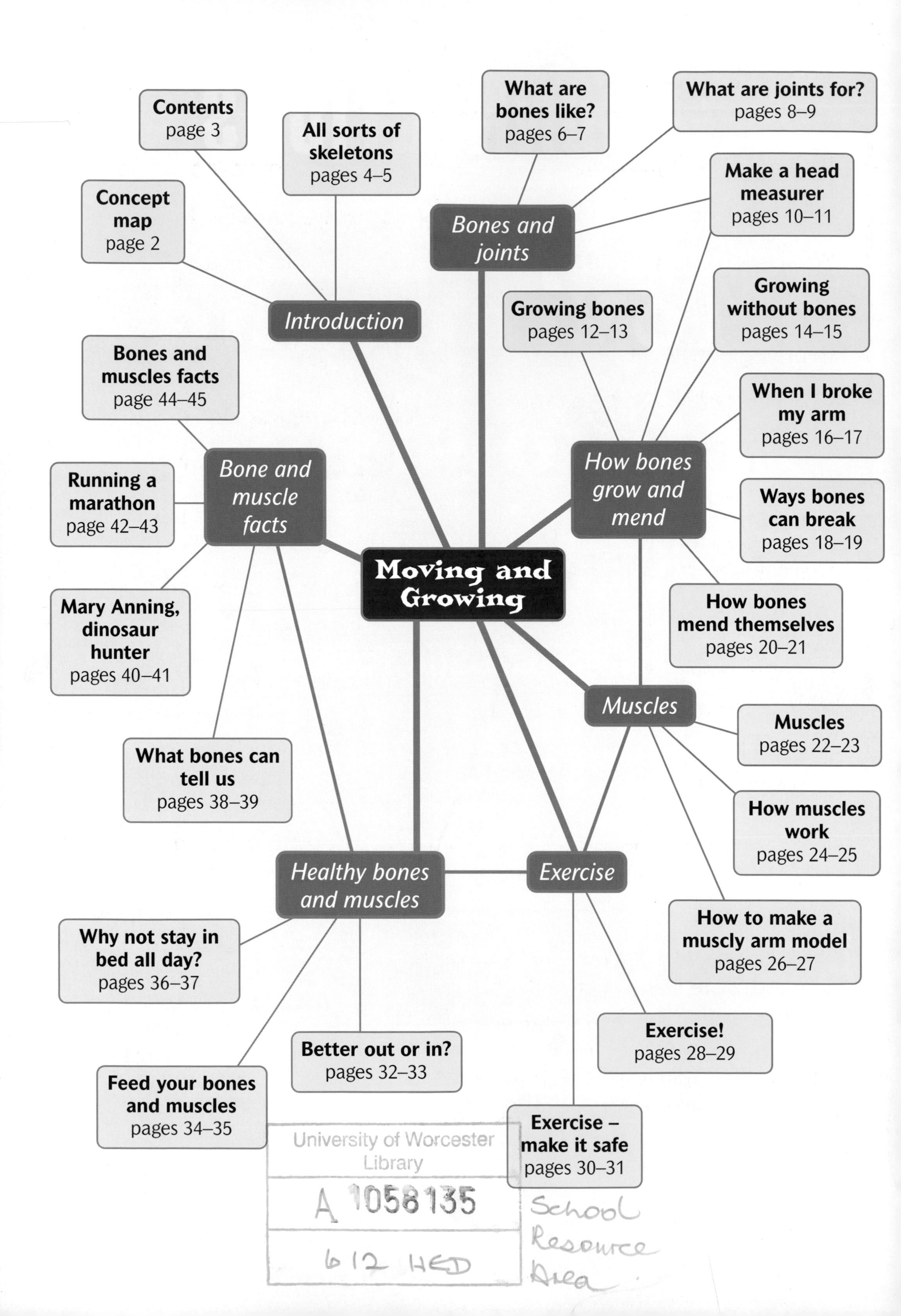
Moving and Growing
Introduction
Concept map page 2
Contents page 3
All sorts of skeletons pages 4–5
Bones and joints
What are bones like? pages 6–7
What are joints for? pages 8–9
Make a head measurer pages 10–11
How bones grow and mend
Growing bones pages 12–13
Growing without bones pages 14–15
When I broke my arm pages 16–17
Ways bones can break pages 18–19
How bones mend themselves pages 20–21
Muscles
Muscles pages 22–23
How muscles work pages 24–25
How to make a muscly arm model pages 26–27
Exercise
Exercise! pages 28–29
Exercise – make it safe pages 30–31
Healthy bones and muscles
Better out or in? pages 32–33
Feed your bones and muscles pages 34–35
Why not stay in bed all day? pages 36–37
Bone and muscle facts
What bones can tell us pages 38–39
Mary Anning, dinosaur hunter pages 40–41
Running a marathon page 42–43
Bones and muscles facts page 44–45

Contents

All sorts of skeletons 4
What are bones like? 6
What are joints for? 8
Make a head measurer 10
Growing bones 12
Growing without bones 14
When I broke my arm 16
Ways bones can break 18
How bones mend themselves 20
Muscles 22
How muscles work 24
How to make a muscly arm model 26
Exercise! 28
Exercise – make it safe 30
Better out or in? 32
Feed your bones and muscles 34
Why not stay in bed all day? 36
What bones can tell us 38
Mary Anning, dinosaur hunter 40
Running a marathon 42
Bones and muscles facts 44
Glossary 46
Bibliography 47
Index 48

Read this book and find out what your skeleton and muscles are made of and how you move. Learn how athletes train for marathons, and what it feels like to break a bone. I especially liked the part that explains how experts use bones to show them the way people lived hundreds of years ago.

Dr Mike Kent
Head of Zoological Programmes,
St Austell College

All sorts of skeletons

A skeleton is a hard framework which holds a creature's body together. It can be **internal**, like a human skeleton. Or it can be **external**, like scaffolding or armour: a crab's skeleton is like this.

Humans are not the only creatures which have an internal skeleton. Birds, fish and many animals do too. They are all called **vertebrates** because they have vertebrae – a long row of bones down the middle of the back, also called the backbone or spine. Creatures with no backbone are called **invertebrates.**

Vertebrates have many things in common. Their skeletons are usually made out of bone. They have cage-like structures to protect the most important **organs**: ribs around the heart and lungs; and a hollow skull that surrounds the brain. Most vertebrates also have bones to provide the structure for **limbs** (legs and arms or wings).

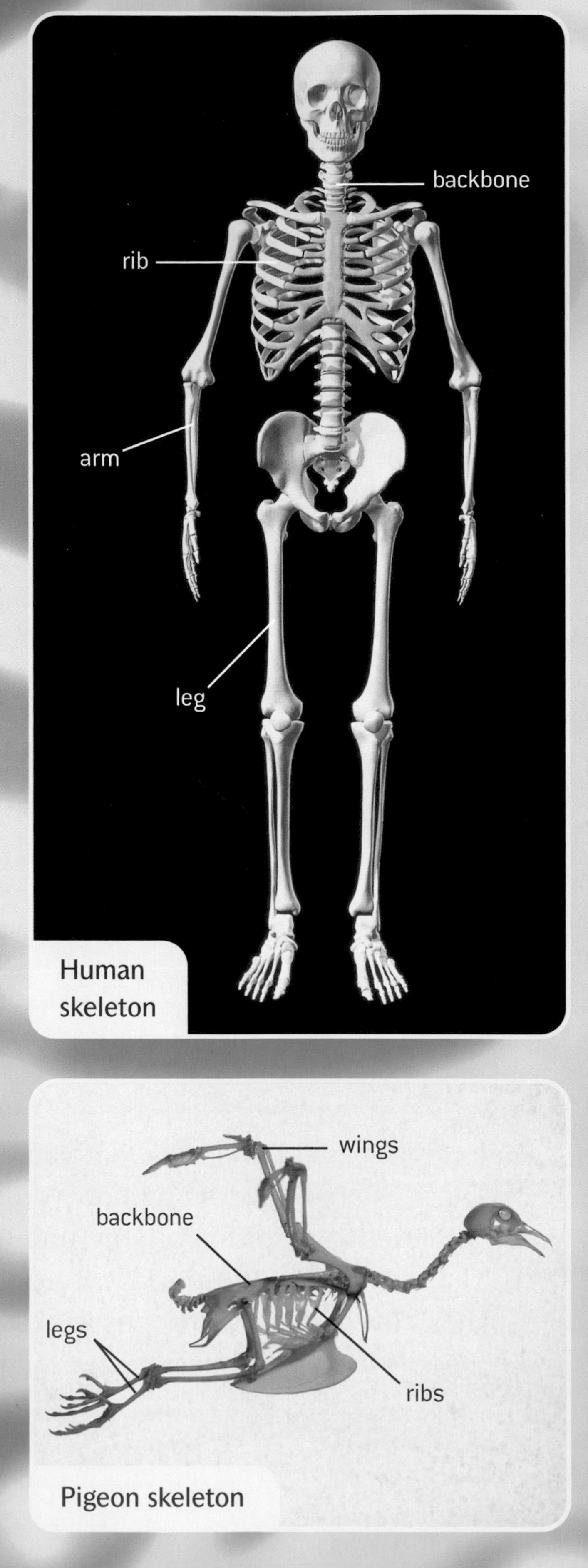

Human skeleton

Pigeon skeleton

There are also some differences between vertebrates. Cats' and dogs' backbones extend along their tails. Some kinds of fish are almost flat, which affects the shape of their ribs. Snakes have no bones for legs and arms. And the size of skeletons is very variable, depending on the size of the animal.

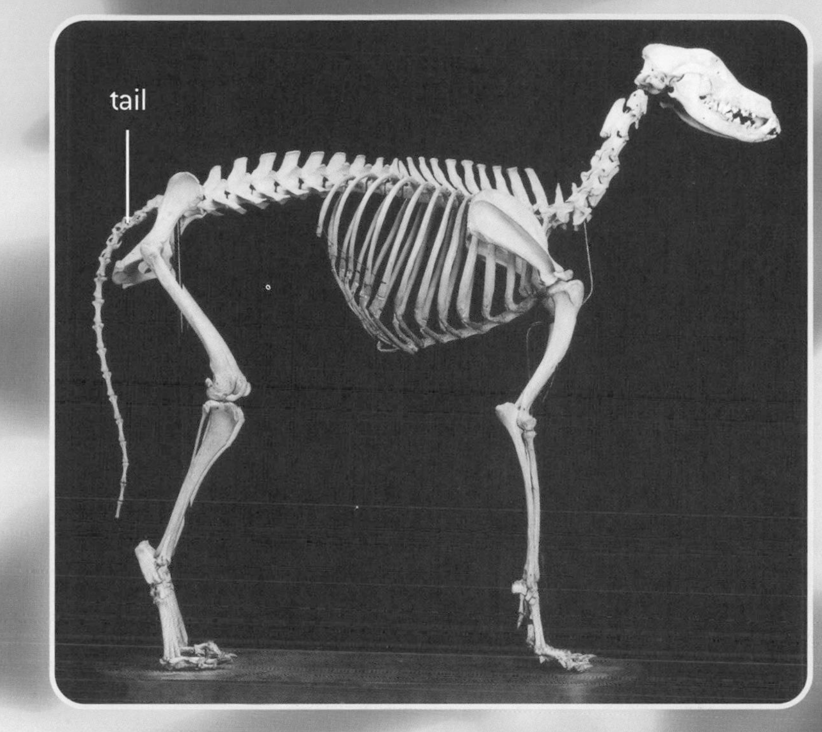

A dog's spine continues along its tail.

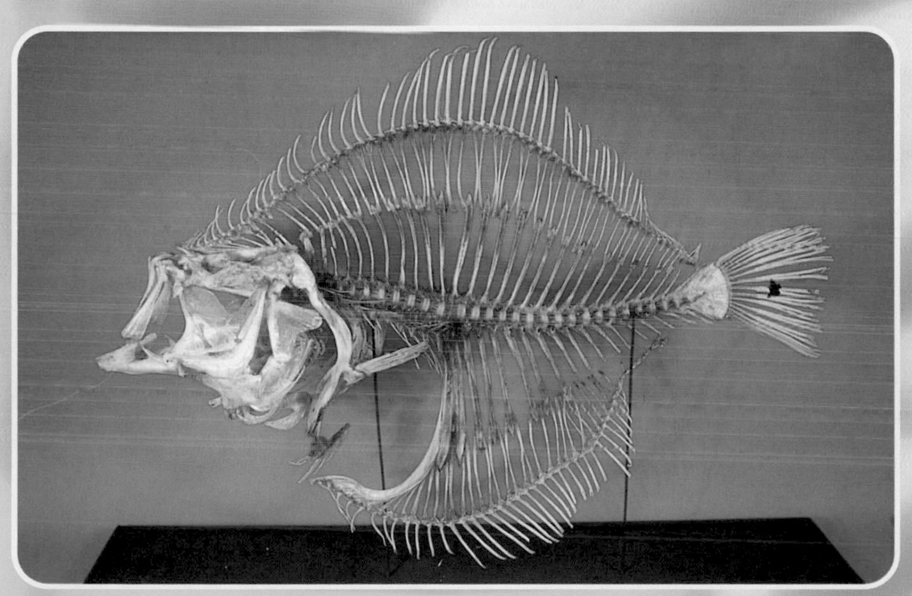

This fish's ribs are flat.

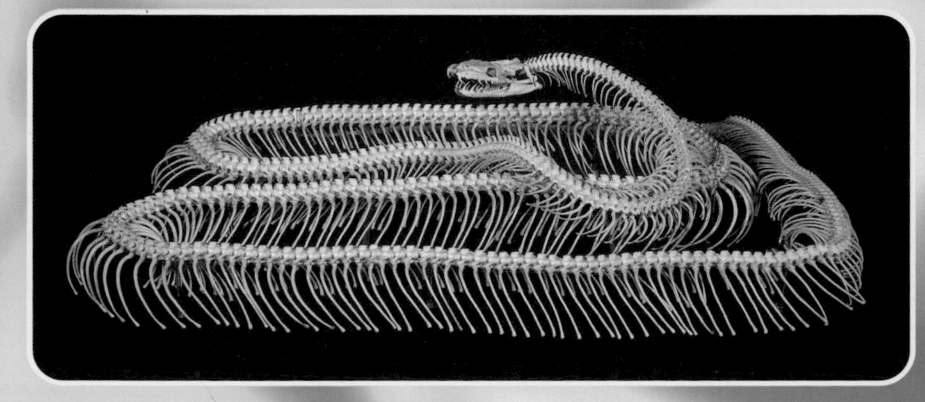

A snake's skeleton has no limbs.

This book will help you to find out more about skeletons, how they work, and how they grow.

What are bones like?

Gently tap your kneecap. Notice how hard it is. Bones need to be hard, in order to be strong enough to hold us up. For the same reason, bones need to be **rigid**, but not too rigid because otherwise they would break easily.

These are the leg bones of a cow. They are large, hard and very thick.

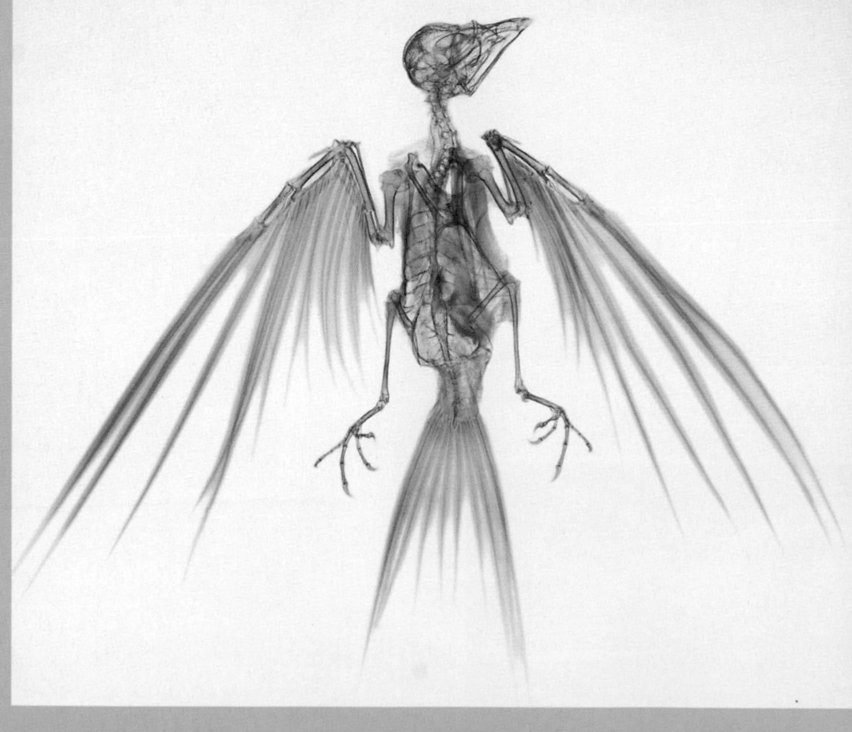

A bird's leg bone is hard, but a lot thinner than a cow's. It is also hollow inside. This is to make it light, so that the bird is able to fly.

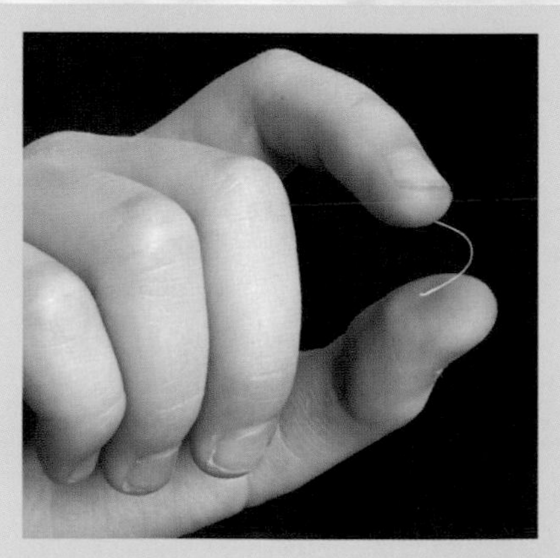

Fish bones are more **flexible** and much finer than animal or bird bones.

A bone is made up of different layers. Each layer has its own job to do.

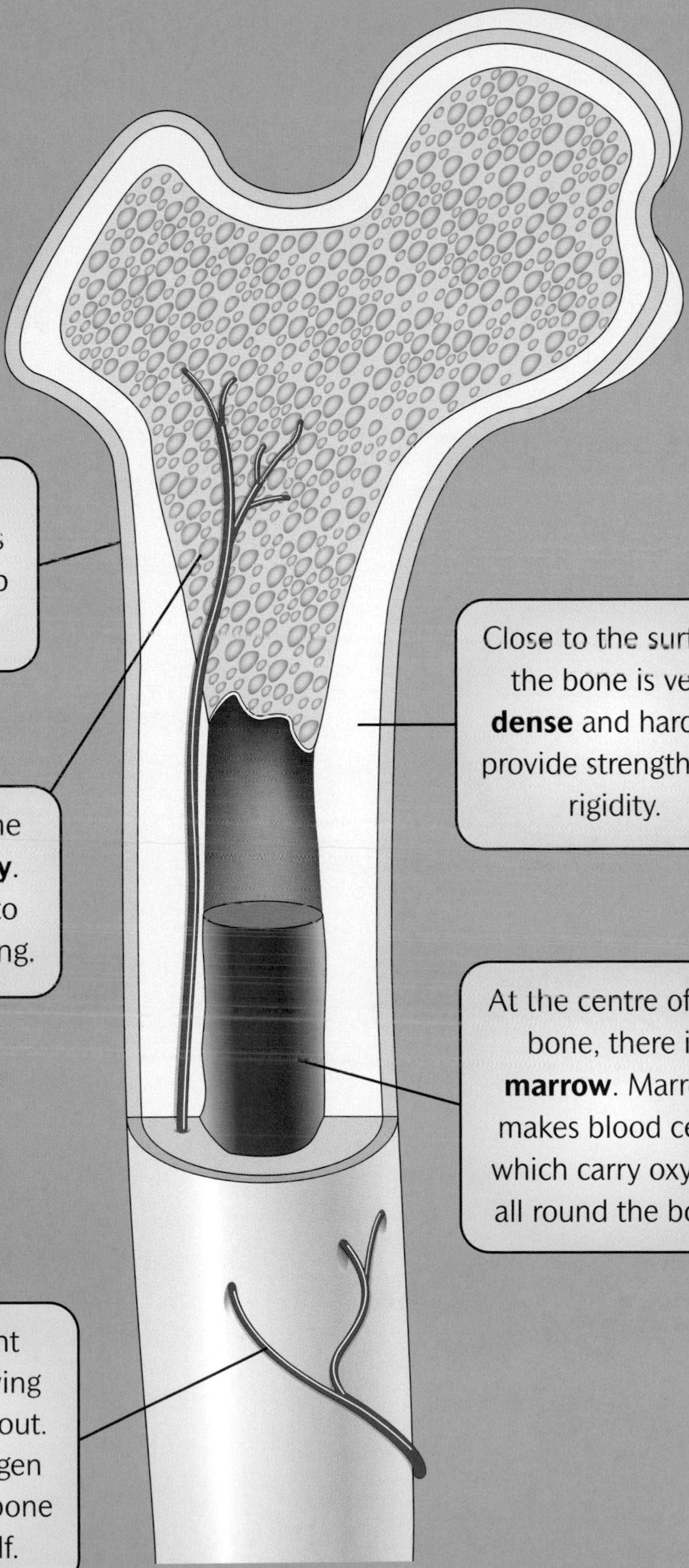

The outside layer is a thin sheath. It contains special cells which help to repair any damage.

Close to the surface, the bone is very **dense** and hard, to provide strength and rigidity.

Near the middle, some of the bone is **spongy**. This helps the bone to be light as well as strong.

At the centre of the bone, there is **marrow**. Marrow makes blood cells, which carry oxygen all round the body.

Blood vessels go right inside the bone, allowing blood to move in and out. The blood carries oxygen and food, to help the bone grow and mend itself.

What are joints for?

Bones have to be strong, hard and **rigid** to hold us up. But we need to be able to bend our bodies too, so that we can move easily. This is why skeletons are jointed.

Joints are the places where one bone ends and another starts. Here the bones can be moved in one or more directions.

It is important that bones only move in the direction they are meant to. For example, our knees must not bend to the side or to the front. So there are various sorts of joints, each one allowing the bones to move in certain ways.

Hinge joints

The finger joints are hinge joints, like the elbow and knee. They work just like a hinge on a door: your fingers can only bend in one direction.

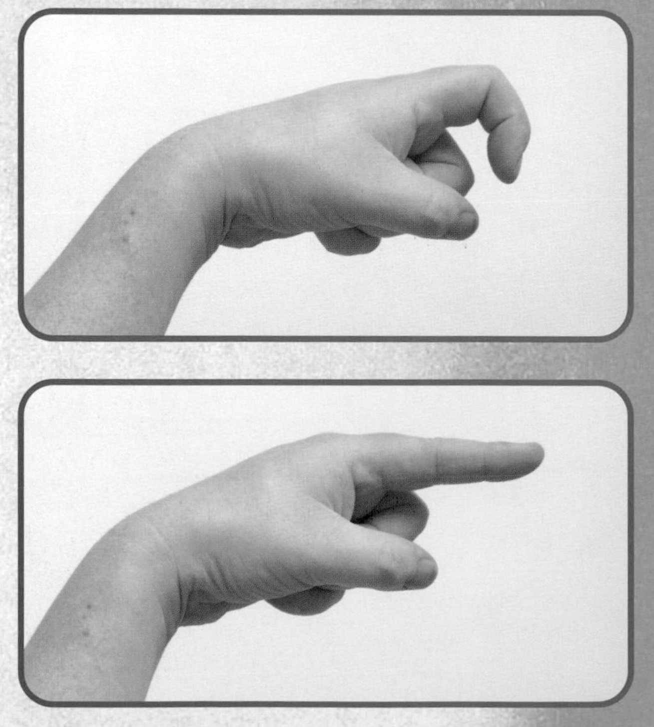

The finger bends and straightens in one direction.

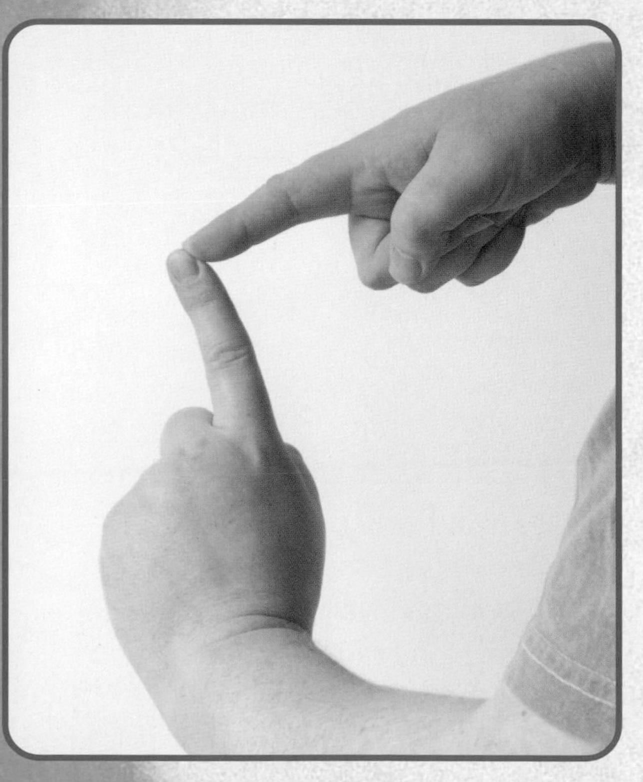

The joint stops the finger bending sideways.

Ball and socket joints

The hip allows movement in all directions – it is very **flexible**.

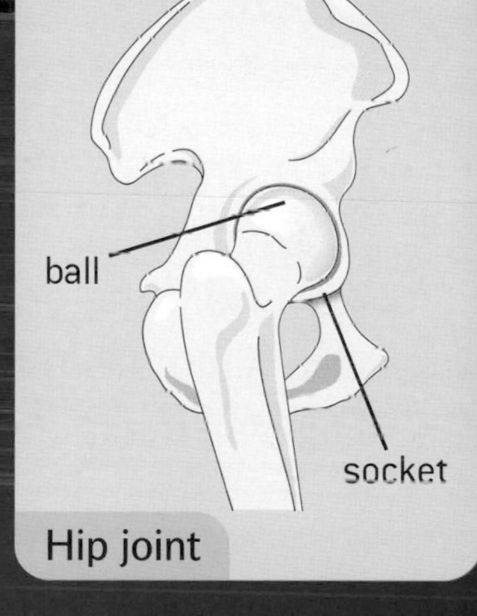

The ball at the end of the thigh bone fits into a cup-shaped socket and allows movement in all directions.

The shoulder joint also allows movement in all directions.

socket
ball
Shoulder joint

Make a head measurer

You will need:

- a long strip of paper
- scissors
- ruler
- pencil
- tape measure

1 Take one end of the strip and fold lengthwise.

2 With the scissors, cut downwards from the fold 2 cm from the end – but stop 5 mm from the edge. This will make a slit.

3 Measure 40 cm away from the slit and make a cross with a pencil, 2 cm in from the edge.

4 Draw a line with a ruler from the cross to the end of the strip, in the opposite direction to the slit.

5 Cut along the line until you get to the cross. Cut down 2 cm to the edge of the strip.

6 Feed the narrow end of the strip through the slit.

7 Place around a partner's head and, using a pencil, mark off their head measurement.

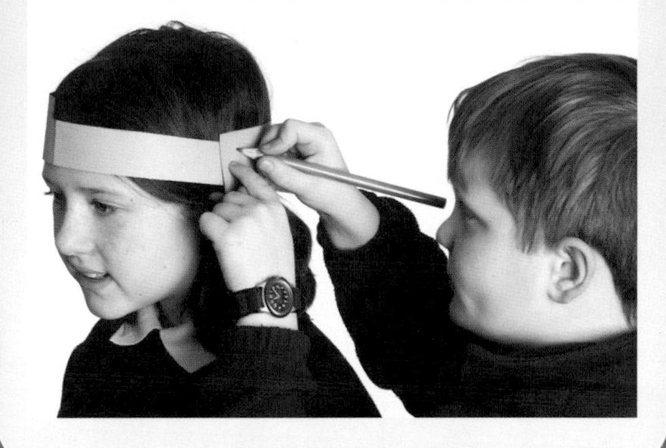

8 Use a tape measure to measure the distance from the slit to the pencil mark.

9 Record your results to the nearest half centimetre on a table like this.

Name	Age	Head **circumference**

Growing bones

Bones grow at different times in our lives and at different rates. One class tested this statement by carrying out a measuring investigation.

Objective

To measure the leg length and head **circumference** of children in our class, plus the teacher and teaching assistants, and to record the results.

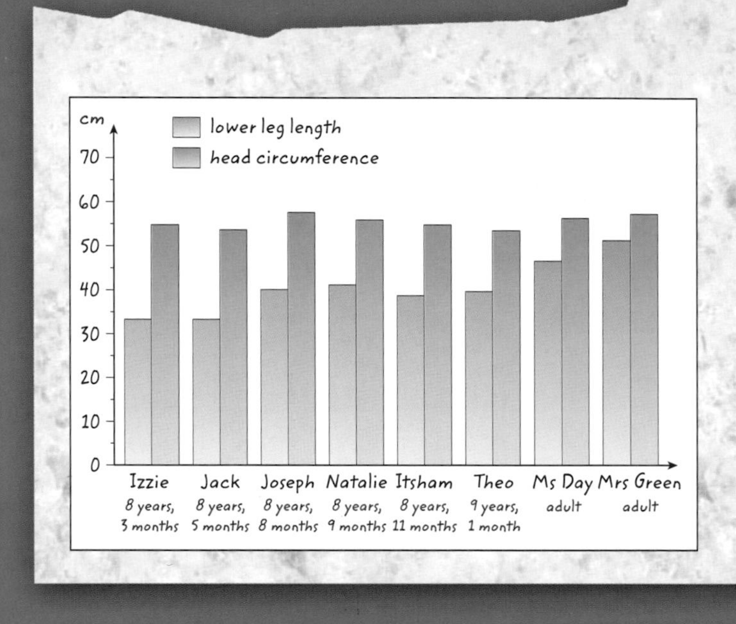

What we did

We used tape measures to measure the length of people's lower leg, from knee to heel. We used a head measurer to measure head circumference. We made each measurement three times, and took the middle answer, if the answers were different. We drew a table and a graph to record the results.

Hats and jackets

As the body grows, the **proportion** of head size to limb length changes.

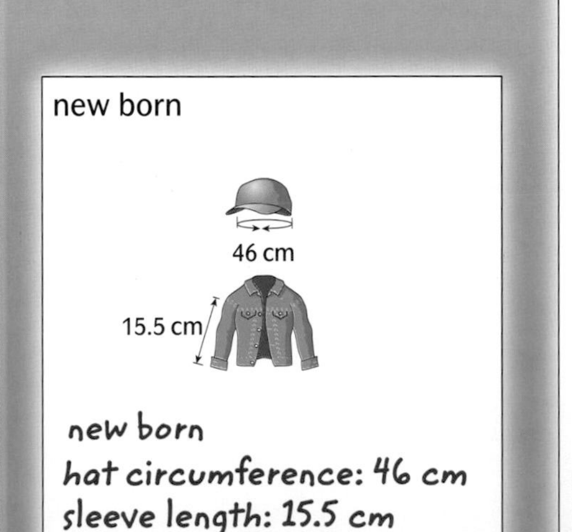

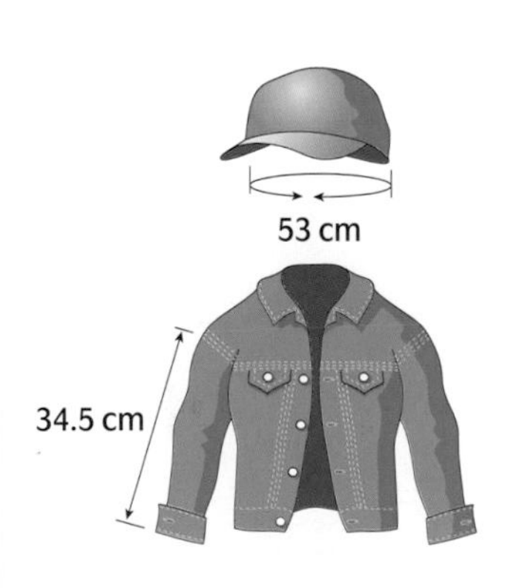

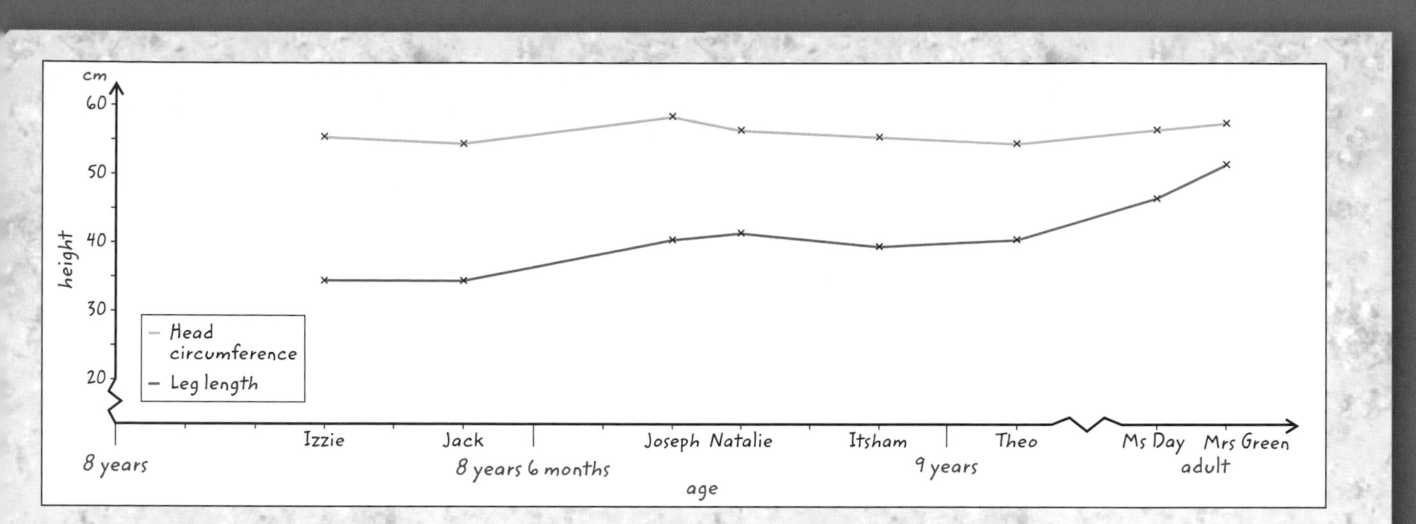

What we found

Name	Age	Lower leg length	Head circumference
Joseph	8 years 8 months	40 cm	58 cm
Natalie	8 years 9 months	41 cm	56 cm
Itsham	8 years 11 months	39 cm	55 cm
Jack	8 years 5 months	34 cm	54 cm
Izzie	8 years 3 months	34 cm	55 cm
Theo	9 years 1 month	40 cm	54 cm
Ms Day	adult	46 cm	56 cm
Mrs Green	adult	51 cm	57 cm

Conclusion

What we found out was that the head does not grow very much after 9 years old, but the lower leg bone does. We also discovered that some older Year 4s had shorter leg bones than some younger Year 4s. This shows that people grow at different rates and that we are not all the same size.

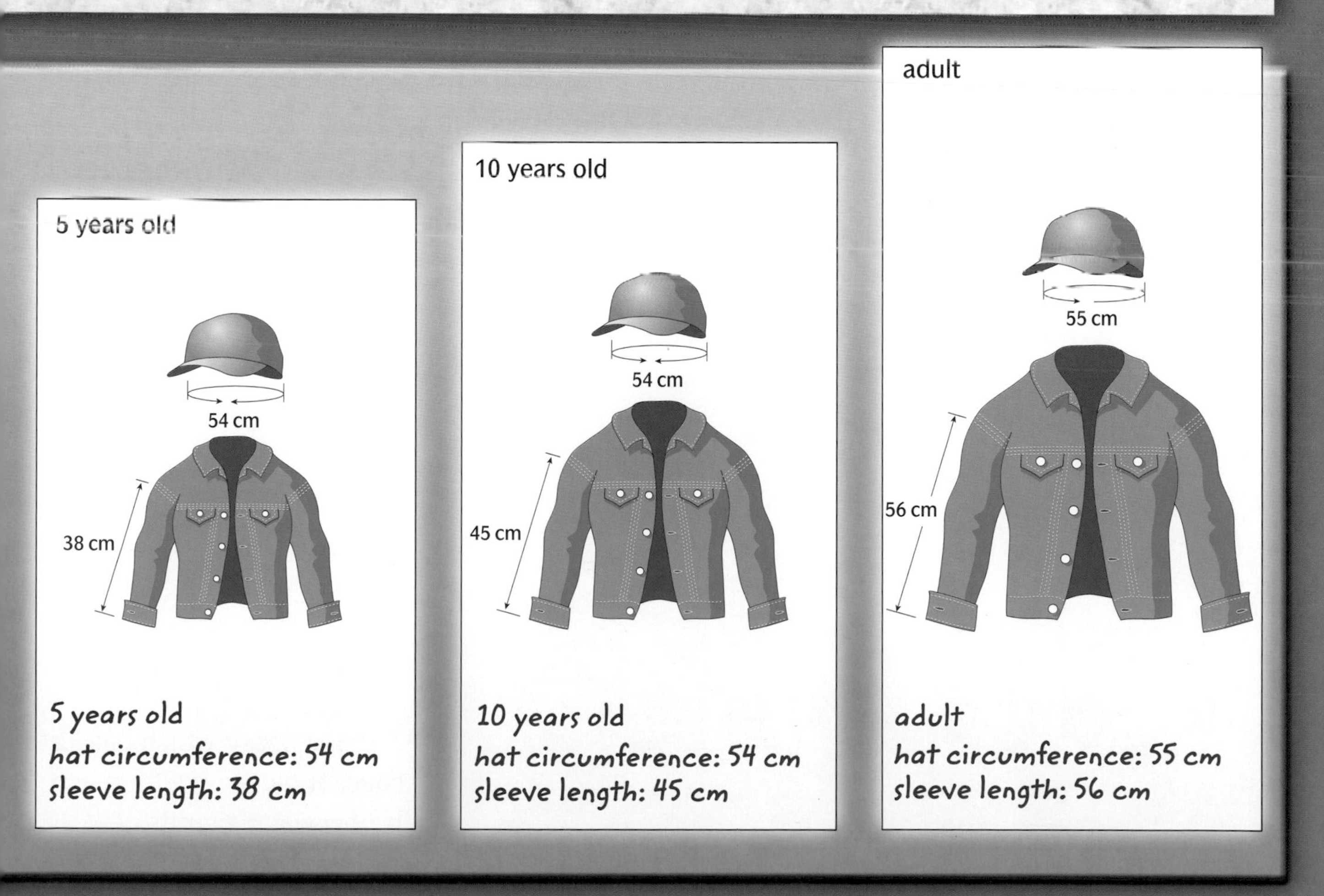

5 years old
hat circumference: 54 cm
sleeve length: 38 cm

10 years old
hat circumference: 54 cm
sleeve length: 45 cm

adult
hat circumference: 55 cm
sleeve length: 56 cm

Growing without bones

Invertebrates are animals that do not have a backbone. Some invertebrates, like worms and slugs, have no bones at all, so they have a very simple shape. Others, like crabs, spiders, butterflies and grasshoppers, have a skeleton on the outside, called an **exoskeleton**.

As a **vertebrate's** skeleton grows, its skin grows too, so a vertebrate gradually becomes bigger until it is fully grown. Invertebrates are more varied than vertebrates, and they grow in a number of different ways, depending on their structure.

Creatures like slugs and worms, which don't have a **rigid** skeleton, simply get bigger. Their skin changes size and gets thicker as they grow.

Invertebrates which have an exoskeleton, like grasshoppers, have a more complicated shape than a worm. But unlike skin, an exoskeleton is rigid – it cannot expand as the grasshopper grows inside it.

When it starts to get too big for its exoskeleton, the grasshopper **sheds** it and stretches out a new, soft exoskeleton from underneath. But this will not be strong until it hardens, which takes time. This is a very dangerous time, as anything could eat the grasshopper while it is weak. So the grasshopper has to sit absolutely still to avoid being noticed and attacked.

A grasshopper which has just shed its old exoskeleton and is now waiting for its new exoskeleton to become rigid.

The life cycle of the butterfly

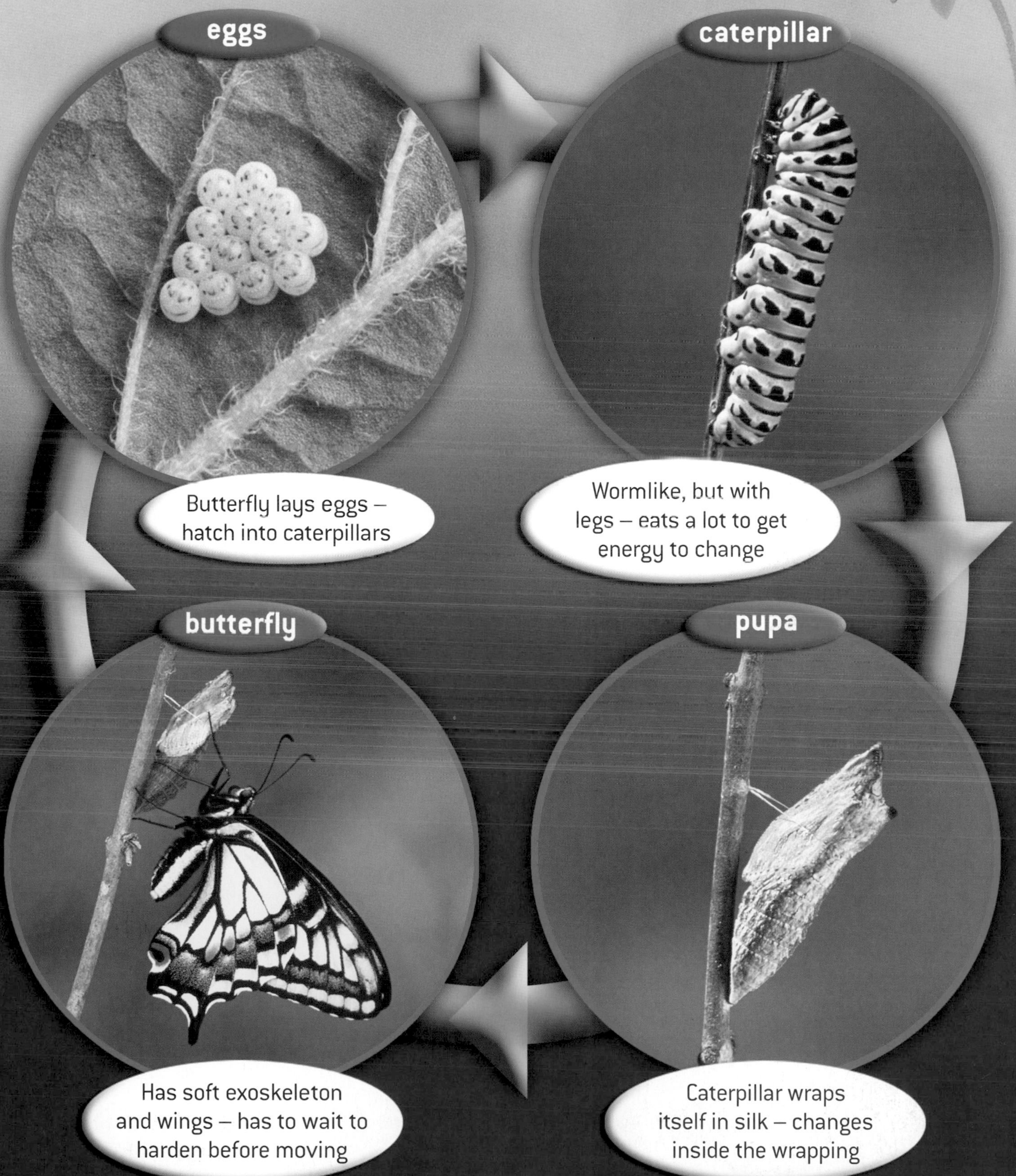

eggs

Butterfly lays eggs – hatch into caterpillars

caterpillar

Wormlike, but with legs – eats a lot to get energy to change

pupa

Caterpillar wraps itself in silk – changes inside the wrapping

butterfly

Has soft exoskeleton and wings – has to wait to harden before moving

When I broke my arm

Last year, I broke my arm, while I was cycling. I fell off and as I landed on my arm, I felt something snap. It hurt so much – it was horrible. I felt very strange and dizzy. Our neighbour saw what happened, and took me home.

Straight away Dad put me in the car and drove to the hospital. It would have been exciting, but my arm hurt very badly and I was scared.

Dad and I only had to wait in the casualty department for about five minutes before a doctor could see us. It still felt like forever. The doctor looked at my arm, gave me some painkillers, and sent me for an x-ray.

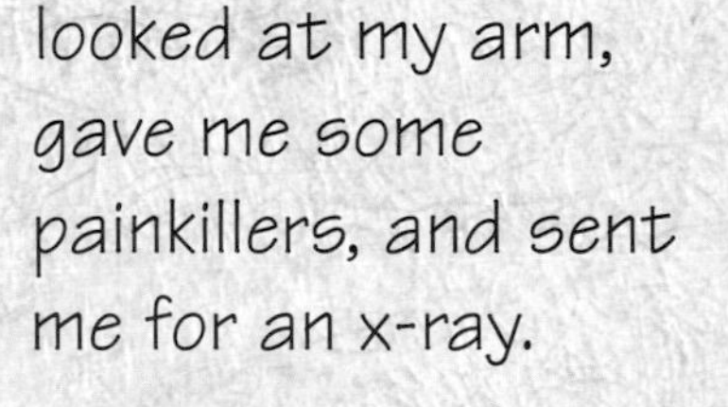

I had to lie very still while the x-ray was taken. When the doctor looked at the x-ray, she told me I had a **fracture**. My arm had to be put in a cast so that the bone would be in the correct position to mend.

For the first two days, I had an open cast, as my arm was quite swollen. Then I went back to hospital to get a complete cast. At school the next day, my friends wrote messages on my plaster, which was fun.

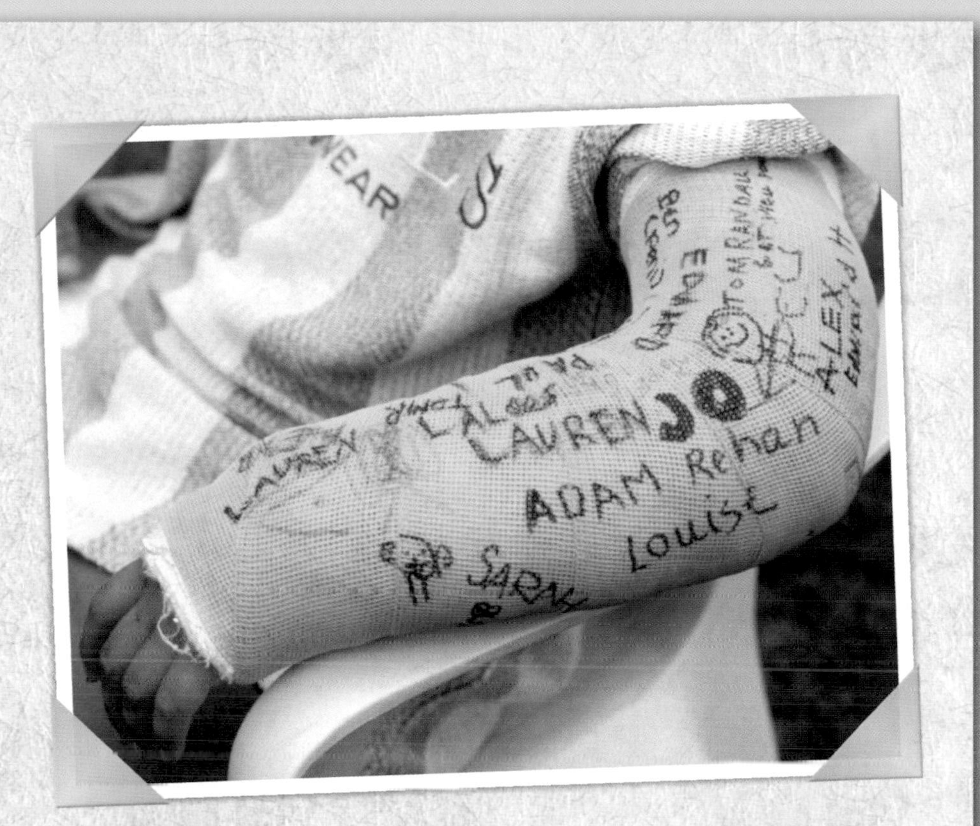

Some things were really annoying. I had to keep the plaster cast dry, so my mum taped a plastic bag over it every time I had a shower. My arm itched a lot under the plaster. It nearly drove me crazy sometimes! I tried scratching the other one, and that helped a bit.

At last the six weeks were up and I could have the plaster taken off. I was so thrilled to have my arm back! At last I was able to have a shower without plastic bags and tape – that was brilliant.

My parents and the doctor say it's safe for me to do all sorts of sports, and ride my bike again. It's exciting – but my arm feels strange without the plaster and I'll be careful not to fall.

Ways bones can break

Although bones are strong, they can break. There is more than one way that bones can break, depending on the person's age and how the accident happened. The medical name for a broken bone is a **fracture**.

Simple fracture

A simple fracture is also called a closed fracture. This is the most straightforward way a bone can break. It just snaps in one place, and the bone doesn't stick out through the skin.

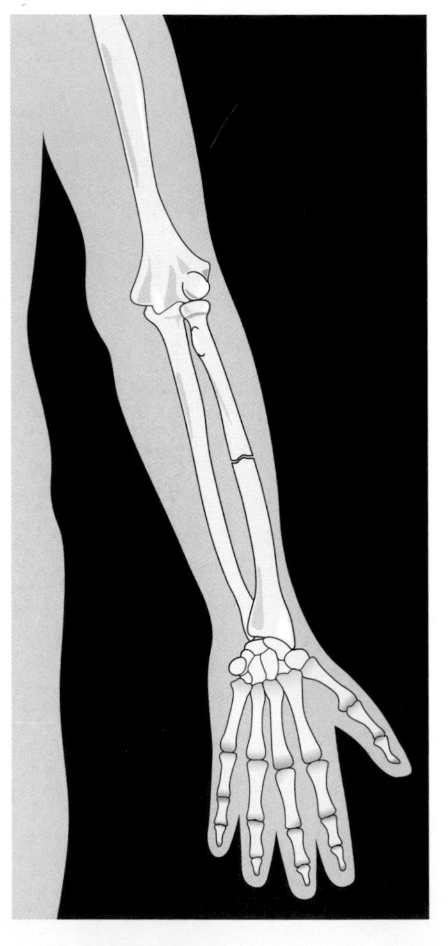

Greenstick fracture

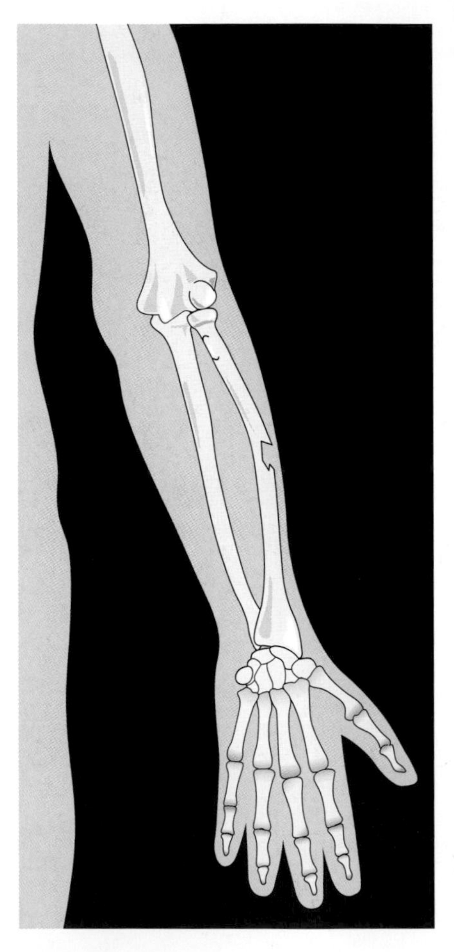

- bone snaps but doesn't break right across
- most common in children – bones more flexible
- 'greenstick' because a green stick breaks in the same way

Compound fracture

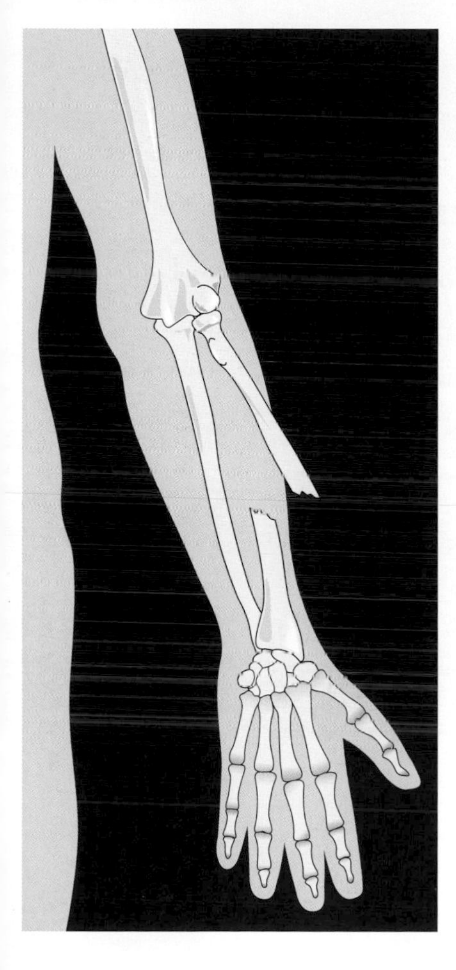

- also called open fracture
- bone end sticks out through skin
- complicated: infection might get in through break in skin

Impacted fracture

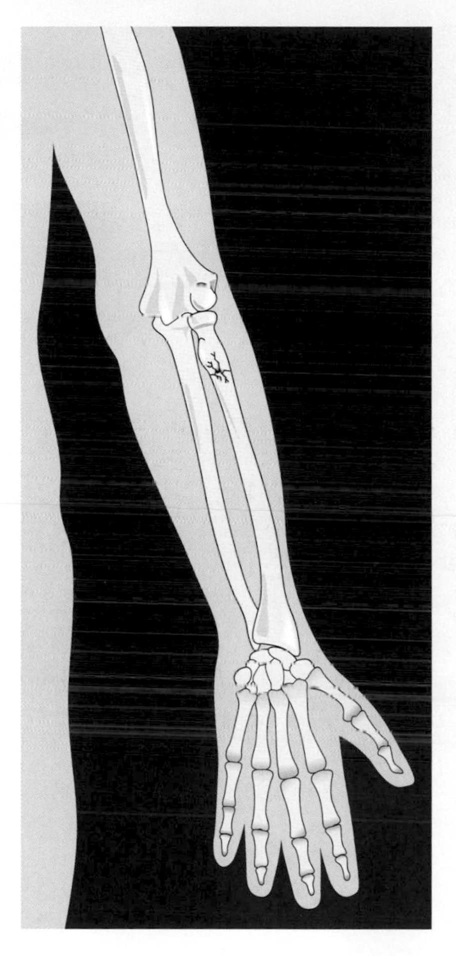

- bone telescopes into itself
- most common in old people

Comminuted fracture

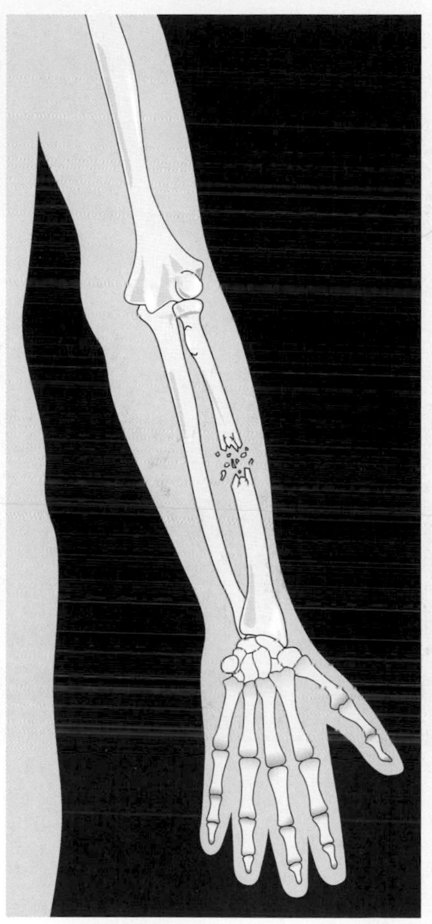

- bone shattered or crushed into lots of small pieces
- most complicated to mend
- pins/plates needed to help mend

How bones mend themselves

When a bone is broken, it takes time to mend itself, going through a number of different stages.

1

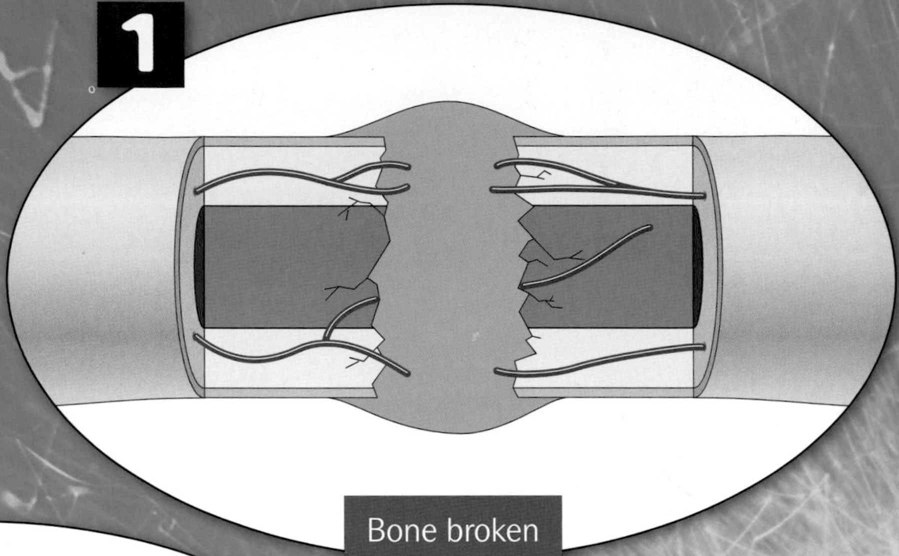

Bone broken

2

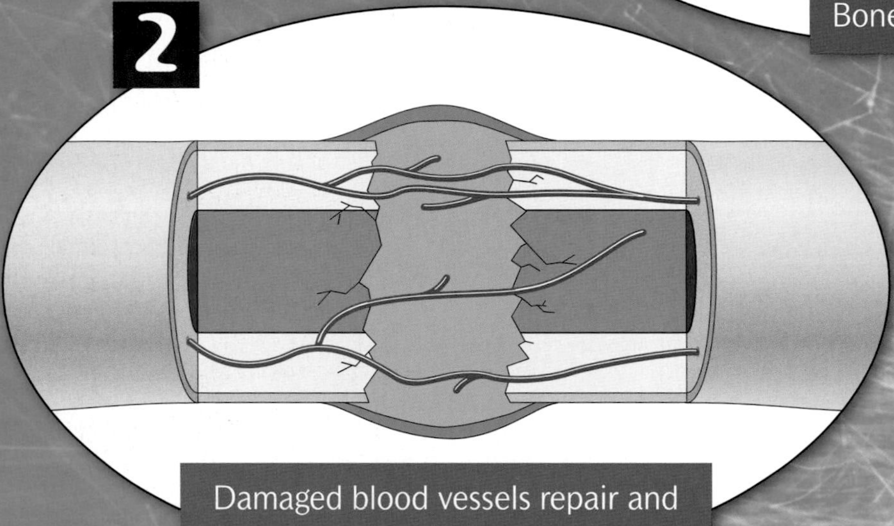

Damaged blood vessels repair and grow into damaged area.

3

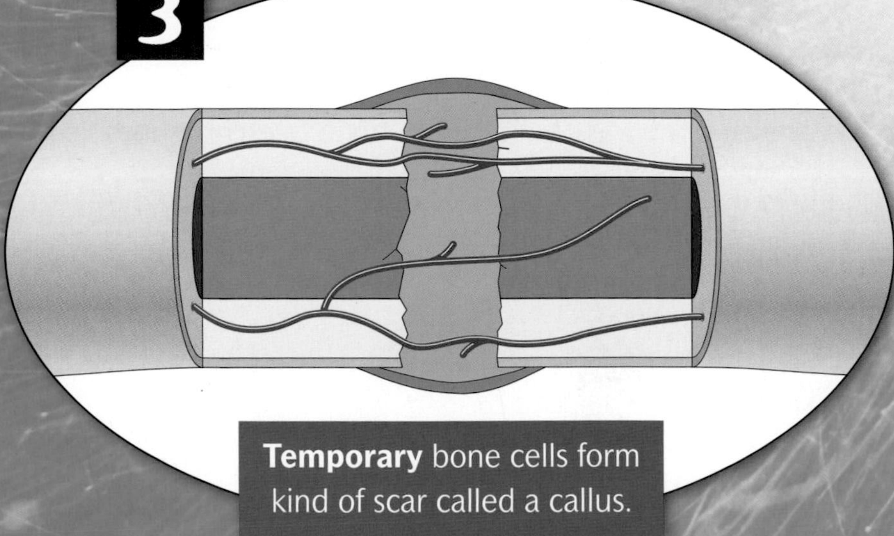

Temporary bone cells form kind of scar called a callus.

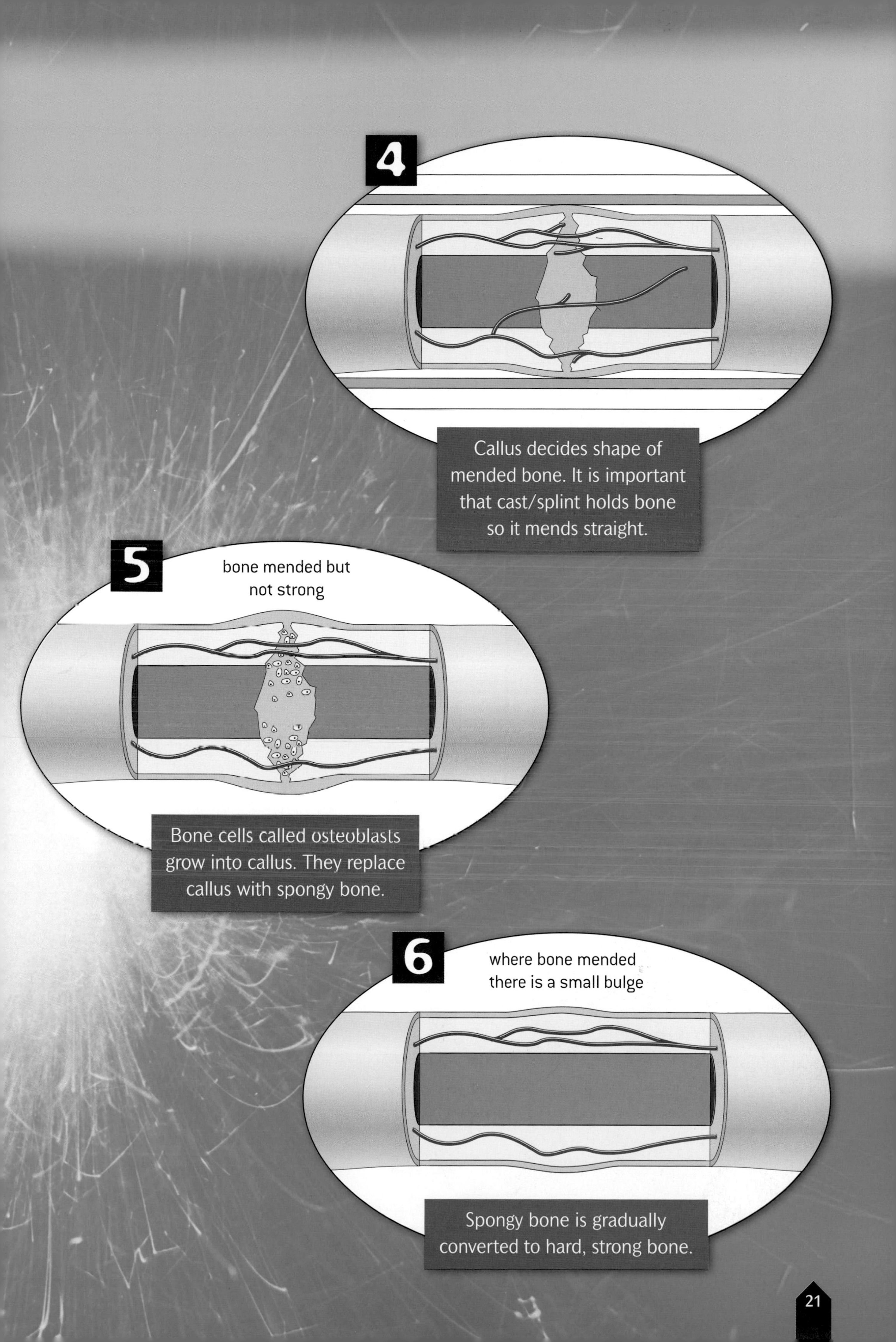
4
Callus decides shape of mended bone. It is important that cast/splint holds bone so it mends straight.
5
bone mended but not strong
Bone cells called osteoblasts grow into callus. They replace callus with spongy bone.
6
where bone mended there is a small bulge
Spongy bone is gradually converted to hard, strong bone.

Muscles

Although the skeleton is jointed to enable the body to bend, bones cannot move on their own. This is one of the reasons why we have muscles. There are muscles all over our bodies, some just under the skin and some further inside our bodies.

Creatures that have no bones, like slugs or worms, are mostly made out of water and muscle. Their muscles do two things: enable them to move and hold their bodies together.

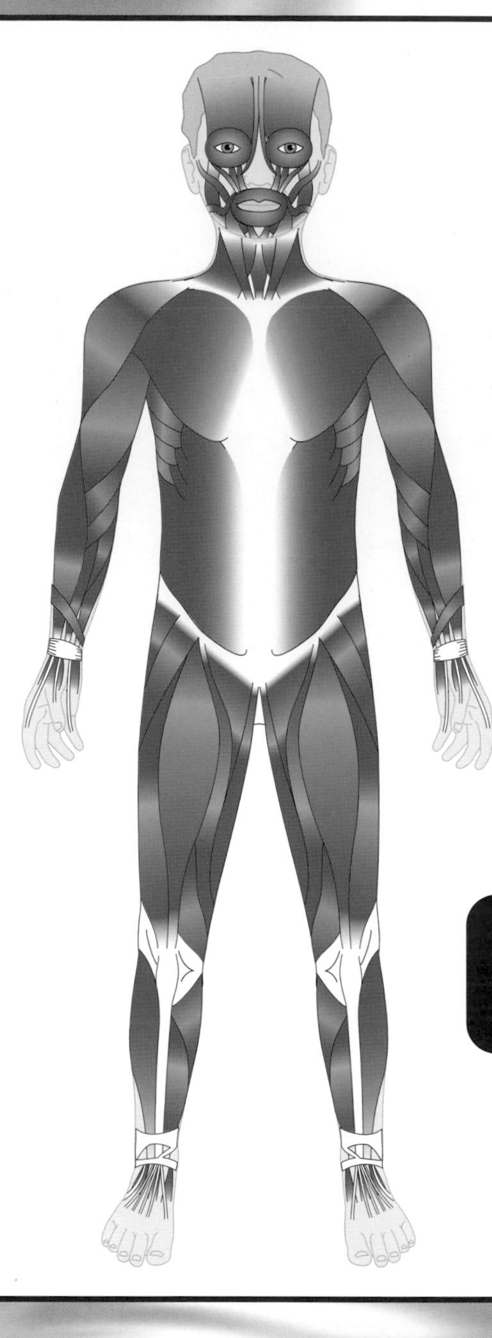

Muscles form a complicated network inside our bodies.

A cross-section through a worm showing the muscles.

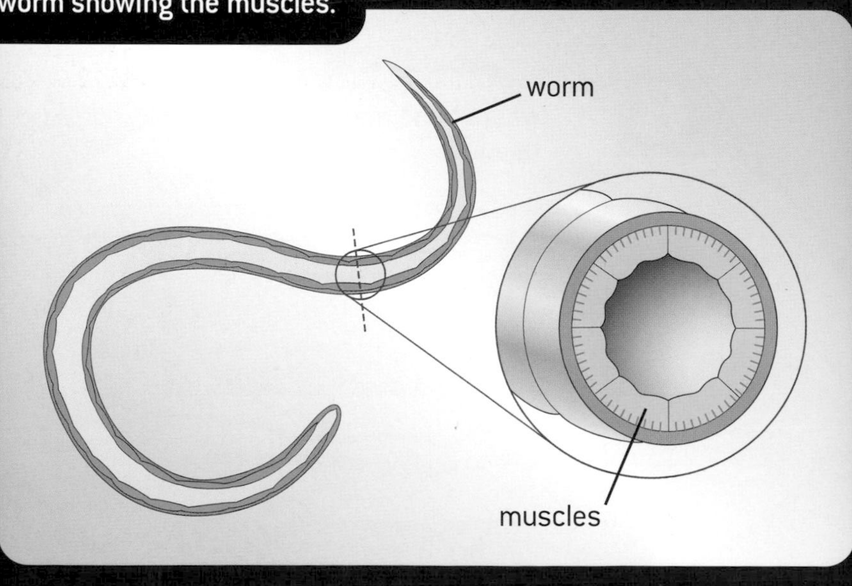

OH, TO BE AN EARTHWORM

Oh, to be an Earthworm.
It has five hearts.
When one is pained or pierced
the other four carry on.
It has no chin to "take it" on,
no upper lip, no backbone
to keep stiff, just crawls
along in closest touch with earth;
doesn't yearn at the stars
or stretch for the moon
but goes about its intimate
business, living its soft life
to the full, savouring it
inch by inch.

Lillian Morrison

Even in humans, muscles make up a very large part of the body. Nearly half of an adult man's body weight is muscle, and about a third of a woman's.

We have three main kinds of muscles: one kind attached to our bones called skeletal muscle, and two other kinds deeper inside our bodies. There is the cardiac muscle which makes up our heart and pumps blood around our body, and the smooth muscle which moves food through our **digestive system**. These 'deep' muscles move automatically – we don't have to think about them moving – and are called involuntary muscles.

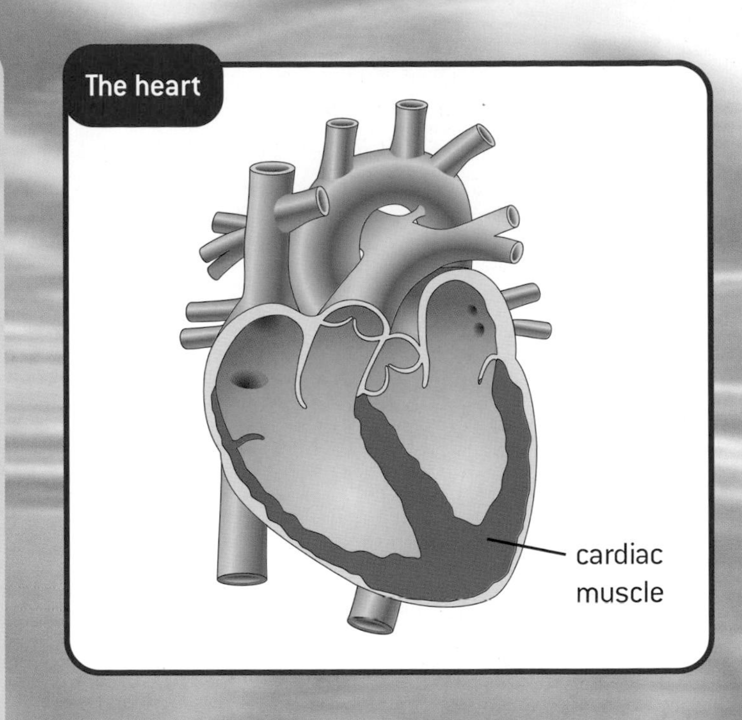

The heart

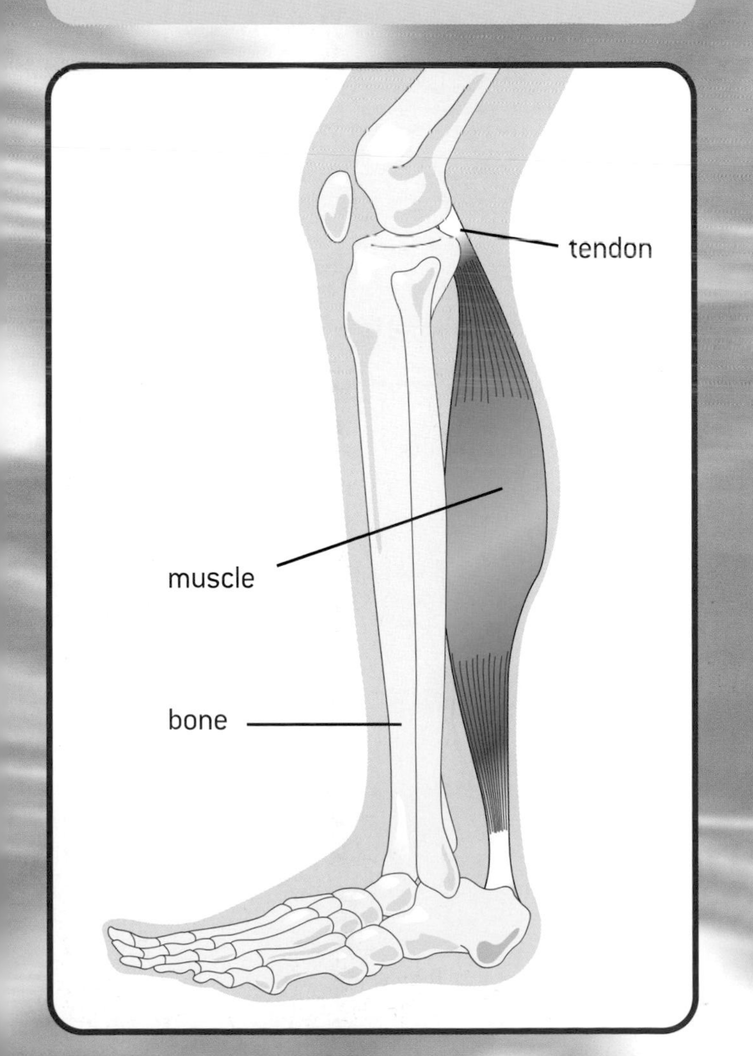

The muscles that we use to move our bones are called voluntary muscles, because they work when we want them to move, rather than moving on their own. These muscles are attached to the bones they move by thick, strong **tendons**, which are a bit like rope.

Whenever you make a movement, like standing up, bending a finger, raising your eyebrows, or breathing, it is your muscles that make it happen.

Meat is the muscle of animals.

How muscles work

The muscles attached to our bones are designed to do two jobs: to shorten(contract), or relax. They are made out of **microscopic** fibres which slide over each other easily when the muscles contract.

When a muscle contracts, it shortens and pulls against a bone, and when it relaxes, it stops pulling. This is all a muscle can do – pull or relax. It cannot push, so muscles have to work in pairs, each pulling in the opposite direction. This is what happens when you bend and straighten your arm.

First, your brain sends a message to your arm muscles, telling the one on the top of your arm (your biceps) to contract. When it does this, it shortens and pulls your lower arm up towards you. This muscle's partner (your triceps) is on the back of your arm – it relaxes when the biceps contracts.

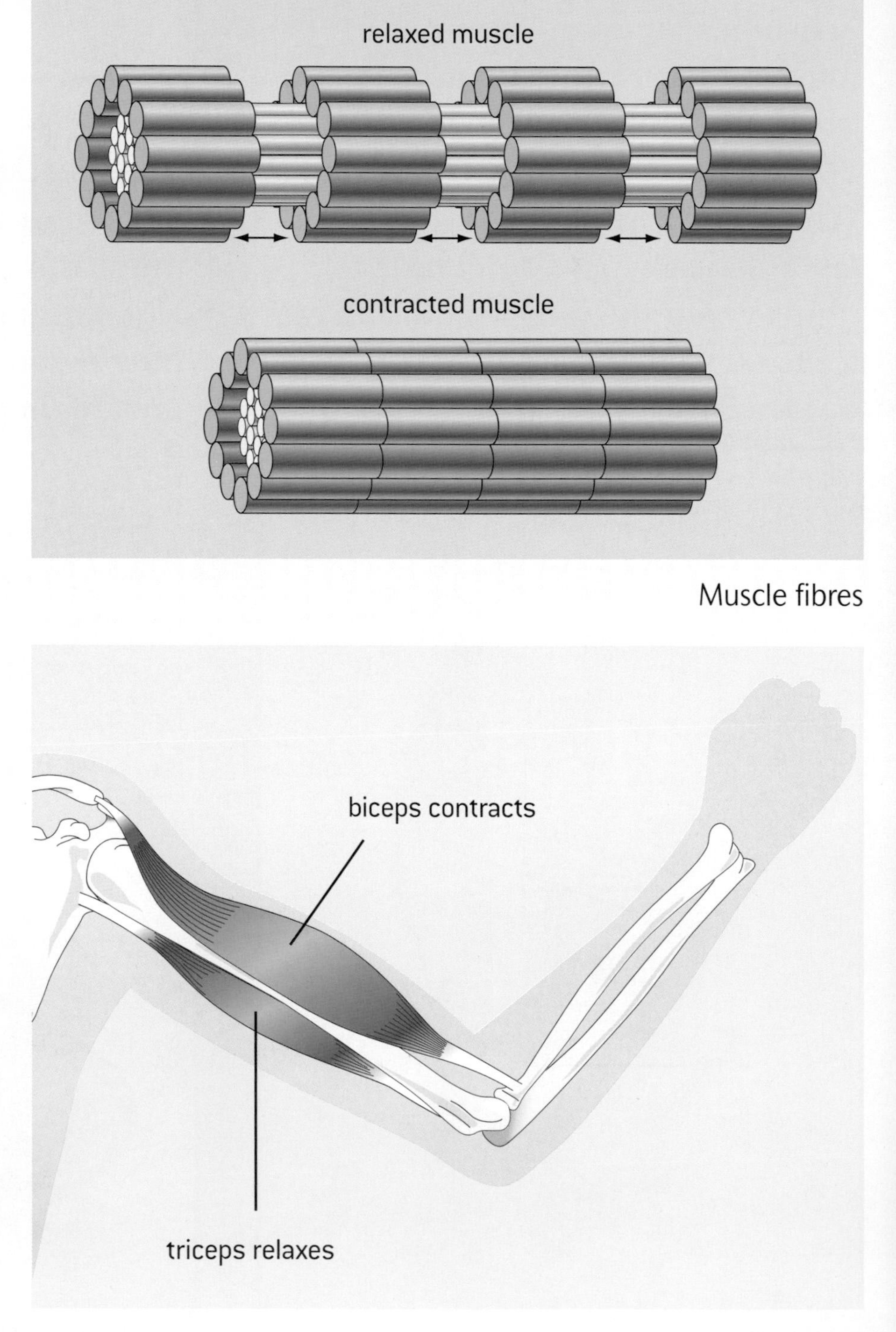

Muscle fibres

Bent arm

If you want to straighten your arm, your brain sends another pair of messages. One tells the biceps to stop pulling. The other tells the triceps to contract. As your triceps shortens, your lower arm straightens out.

You can feel this happening in your arm as you bend and straighten it. The muscle on the top gets harder as you bend the arm, and the one underneath gets harder when you straighten it.

It is an effort for muscles to work, which is why you get tired when you exercise. The muscles become hot too – touch your legs and arms next time you have been running around. You can feel how hard your muscles have been working.

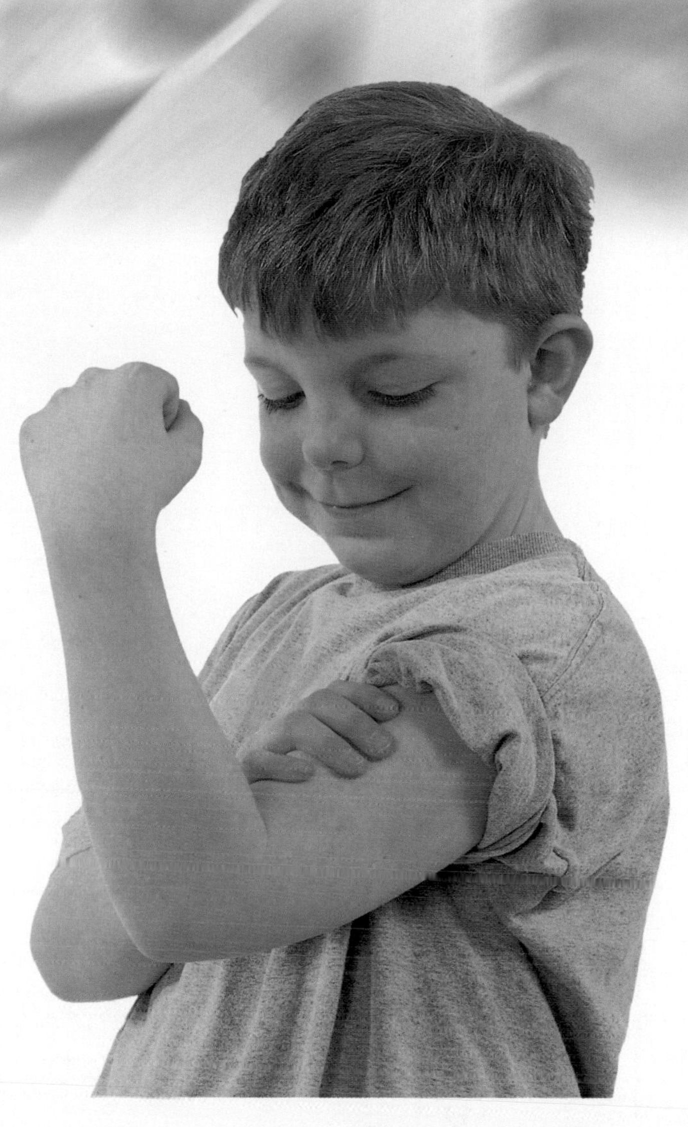

Biceps contracting

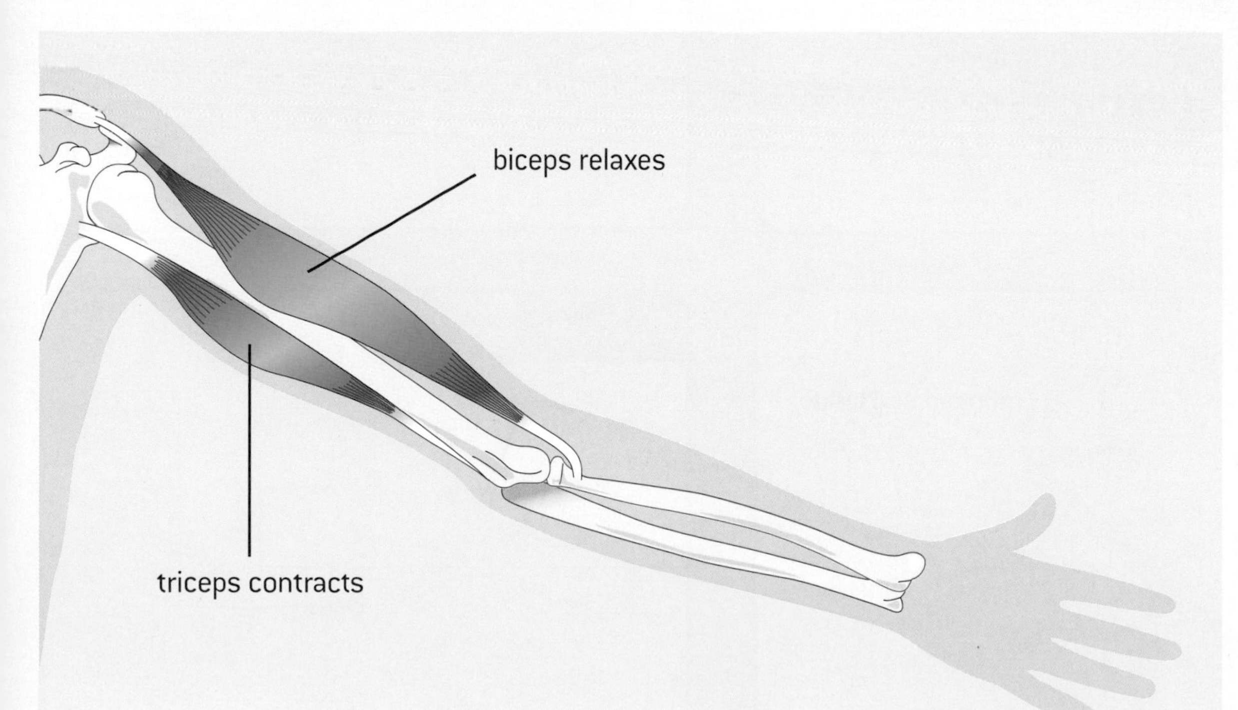

Straight arm

How to make a muscly arm model

Make the model to see how your arm muscles work.

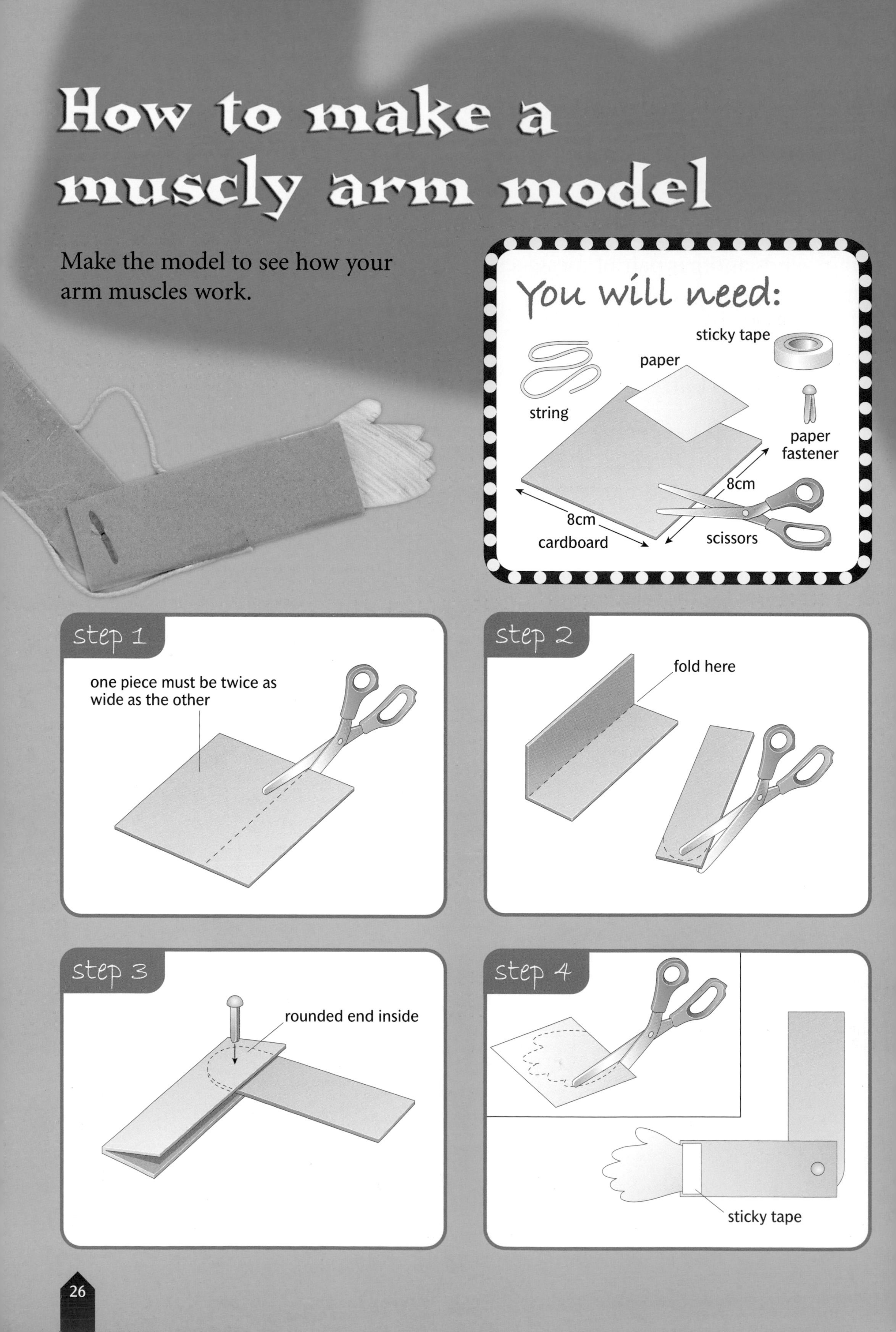

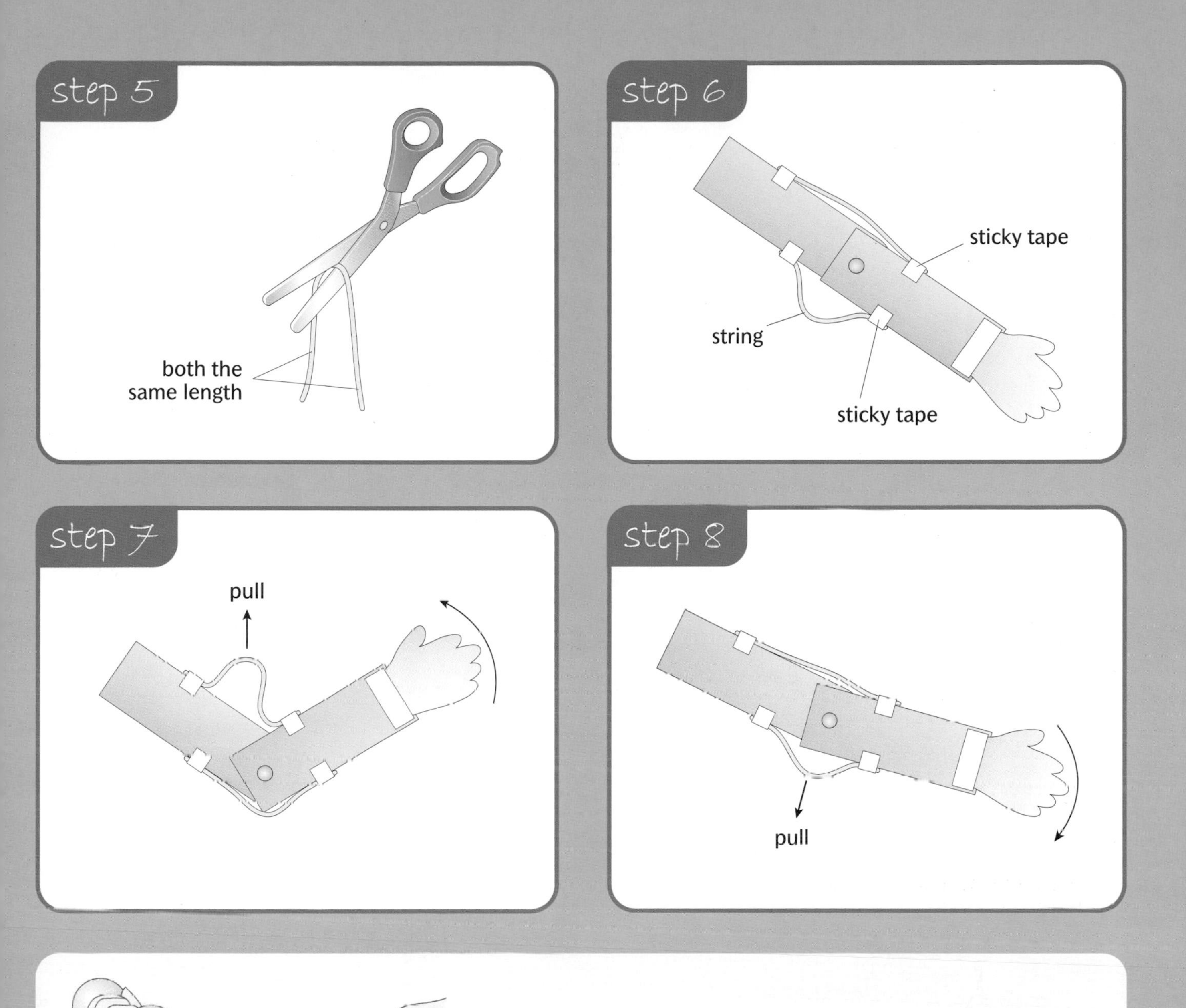

Your muscles bulge as they move the bones from one position to another.

Your model works in the same way as your arm: the bottom part of your arm has two bones and the top part has one. The arm can only bend in one direction.

Exercise!

Re-Flex Health Club

– all the finest fitness facilities under one roof

Are you getting enough?

We all know that exercise is important: doctors say everyone needs at least three half-hour sessions every week. But, be honest, do you get anywhere near this? Most of us don't – and more than half of the adults in this country are overweight.

Do yourself some good!

Scientists have proved that a body which gets regular exercise is a healthy body. Regular activity helps control weight, tones muscles, and even boosts the immune system. If you had a dog, wouldn't you exercise it regularly? Do yourself the same favour – you'll really feel the benefits.

Get fit – feel great!

Exercise can do so much more than improving your body. Stressed? Not sleeping well? A good work-out can burn off your tension, putting an end to sleepless nights. And how about your social life? You can make friends at fitness classes or in the gym – and feeling good about your body will really boost your confidence.

Free taster sessions throughout August!

We've got the lot

Re-Flex has everything you need – a-state-of-the-art fitness room with rowers, treadmills and toning machines, all the latest exercise classes, 25-metre pool and a luxurious sauna suite. And our friendly, highly-qualified staff are always on hand to answer queries or help you devise a personal fitness plan.

Come down to Re-Flex – book for your free taster session now!

We can help you change your life.

Exercise – make it safe

Class 4
Willow School
Willow Close
Willowtown
WW10 9JU

15 October 2003

Dear Mr Gilbert

We are writing to ask permission to hold a special event at the adventure playground on Bourne Road. We want to hold a sponsored exercise-a-thon.

This event will be good for the community – first, it will raise money to buy sports equipment for the school. Secondly, it will encourage more people to use the adventure playground. Thirdly, it will make people think about doing some exercise.

Too many of us don't do enough exercise and risk serious illness because we are overweight. Our parents and teachers are always saying we should be running around outside instead of watching television or playing computer games.

This is a chance to show everyone what else we can do with our free time, and to encourage other children to do the same. We will also be raising funds to improve our school equipment so more children can enjoy sports and games.

Please give your permission for this exciting project.

Yours sincerely
Class 4

Willow School, Willow Close
Willowtown, WW10 9JU
Tel: 01885 23458 email: willowschool@edu.co.uk

17 October 2003

Dear Class 4,

I read with interest your proposal for a sponsored exercise-a-thon at the adventure playground on Bourne Road. I am certainly aware that it is under-used, that getting people to exercise is important, and that extra funds for school sports equipment would be very welcome. However, your suggestion raises a number of concerns.

Any school activity has to be supervised, but there are no staff based at the adventure playground. The school would have to find people to do this job – but this may be difficult as few people are trained to deal with the sort of emergencies that can arise.

You are strong and healthy children, but the sort of activity undertaken in an adventure playground is quite dangerous. Swinging and climbing activities often lead to falls, and there is always a serious risk of straining muscles, and even breaking bones.

However, I agree with your points about raising the profile of exercise and improving the school's sports facilities. So I am contacting the swimming pool about holding a swim-a-thon instead. The pool has well-trained lifeguards in attendance at all times, which answers the main safety concern, and the money raised could be used to buy new sports equipment, as you suggest. If you want to get more people to use the Bourne Road playground, perhaps you could make posters to draw attention to it – I will set aside space on the school noticeboard to put them up.

Yours sincerely

W. Gilbert

Mr W. Gilbert
Head

Better out or in?

Human beings and other **vertebrates** have an **internal** skeleton. But creatures like spiders, grasshoppers and flies, known as **arthropods**, have an exoskeleton. Each design has its own advantages and disadvantages. Which would you rather have?

Ant

Scorpion

Exoskeleton

There are several advantages to having an exoskeleton. First of all, it does not tear or graze like skin and secondly, it can be completely waterproof, so that creatures such as locusts can live in deserts because they do not lose water through **evaporation**. Thirdly, for very small creatures, an exoskeleton is light and very strong.

As with vertebrates, the exoskeleton protects the creature's internal organs. The shape of an invertebrate is defined by its exoskeleton. As 85% of creatures have an exoskeleton it would seem that an exoskeleton can adapt itself to a wide variety of shapes.

However, there are also disadvantages to having an exoskeleton. It is not sensitive like skin, so arthropods need special sense organs to feel. The exoskeleton is made of **chitin** which is very strong, but an arthropod as big as a person would need an exoskeleton so thick it would be too heavy to stand – this is why arthropods tend to be no bigger than a human hand.

Seahorse

Another disadvantage is that an arthropods has to shed its exoskeleton in order to grow. When it has shed its exoskeleton, the creature has to sit very still while its new exoskeleton hardens. This makes it very **vulnerable** to predators.

Finally, an arthropod finds it difficult to control body temperature because an exoskeleton does not breathe in the way skin does.

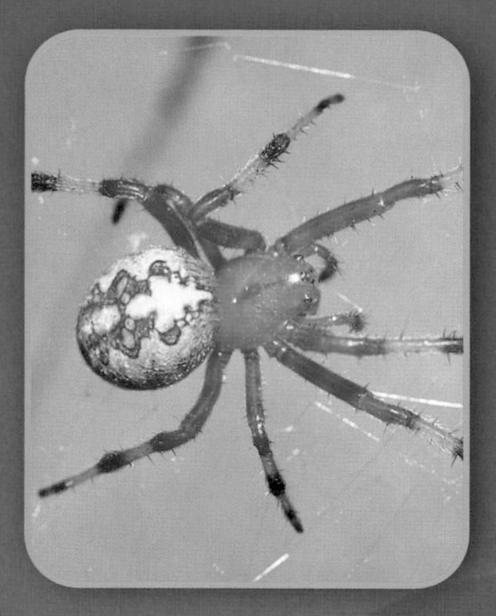

Spider

Internal skeleton

Advantages

- The skeleton defines the creature's shape.
- Skin mends more easily than an exoskeleton, so most **external** damage is swiftly healed.
- Cages of bones protect the most important organs.
- Skin grows with the creature, so it is safer to grow. There is no need to shed an exoskeleton.
- Vertebrates can be bigger than arthropods, because bone is very light for its strength, even in large creatures.
- It is easier to control temperature with skin on the outside – heat can be lost through the skin.
- Skin is sensitive to touch, so animals with skin can feel all over.

Elephant and mouse

Disadvantages

- Skin can be damaged more easily than chitin – but it is also better at mending.
- Most vertebrates would find life in a desert too dry because of evaporation through the skin.

This would never happen!

Feed your bones and muscles

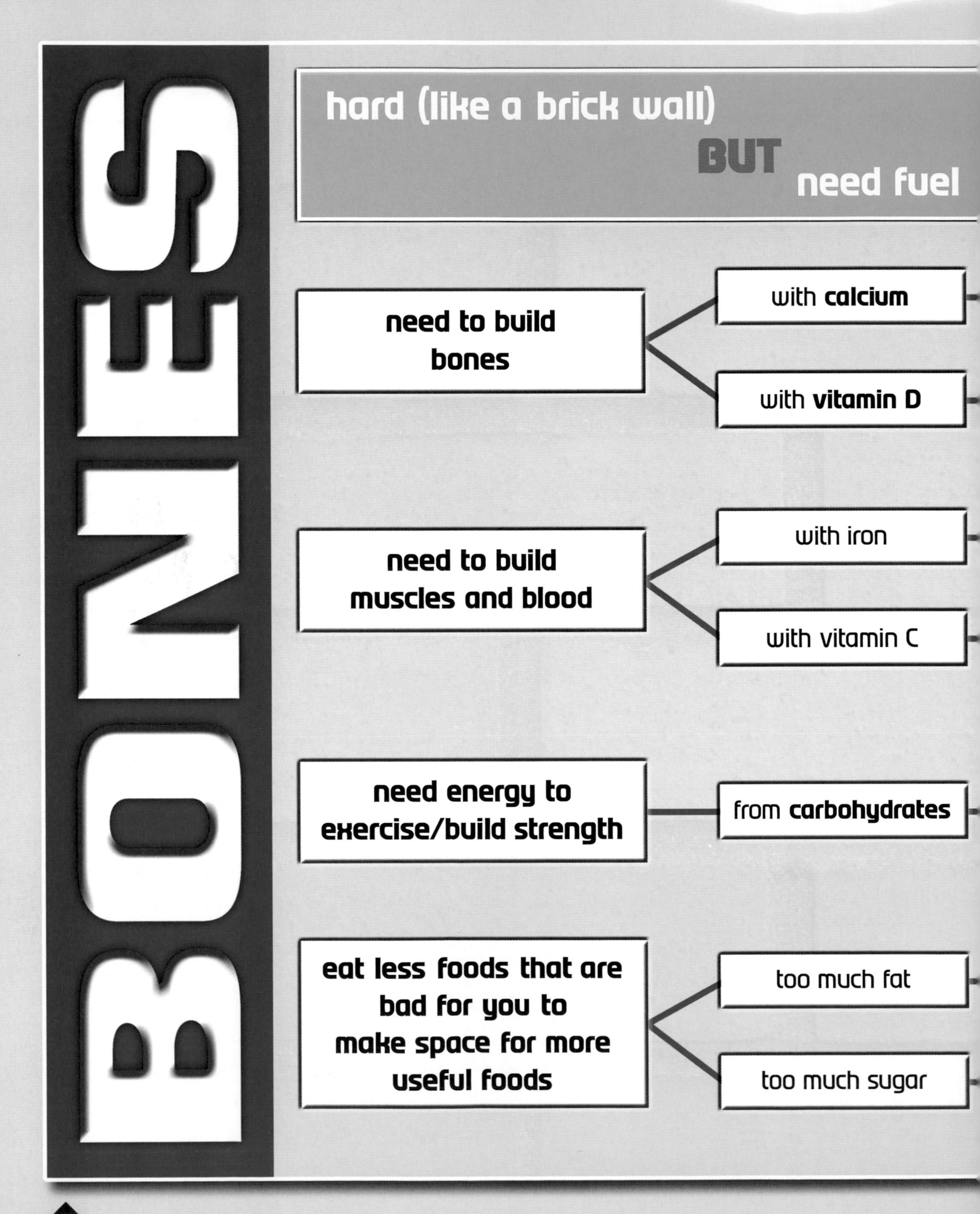

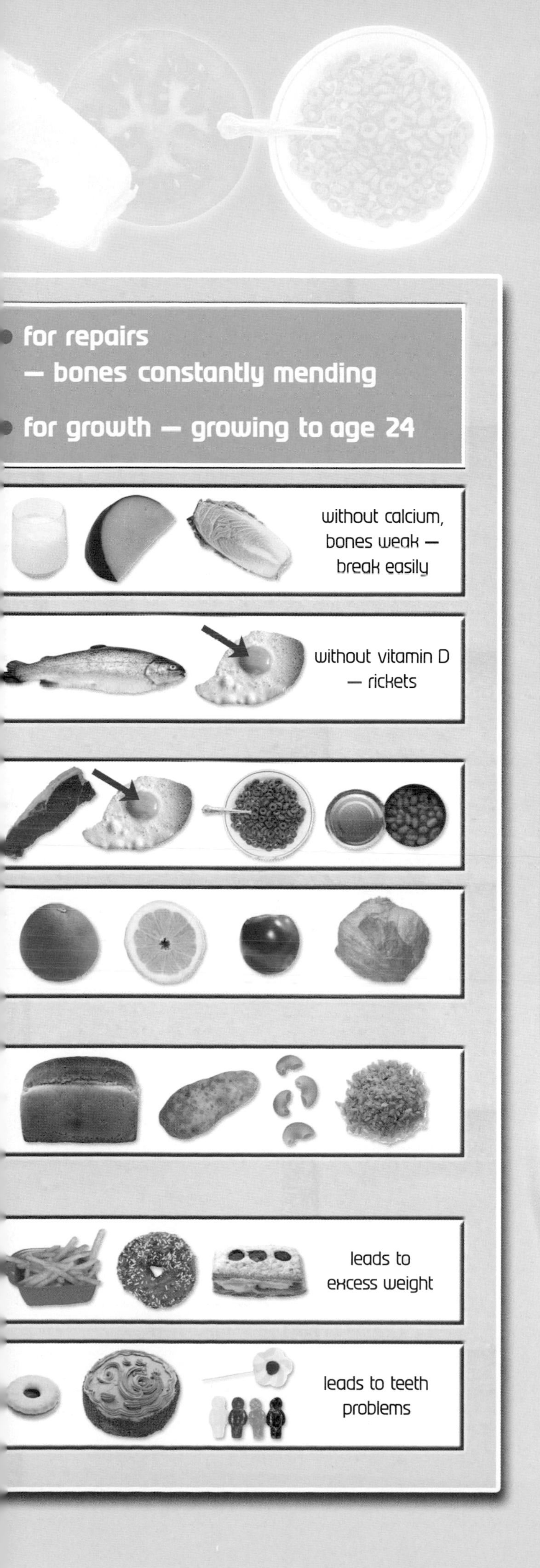

Rickets

This is a condition which affects people, especially children, who do not get enough vitamin D. The problem usually affects people who do not get out in the sunshine, or wear clothes that completely cover their skin. Vitamin D is needed to 'fix' calcium in the bones – without it, the bones are weak.

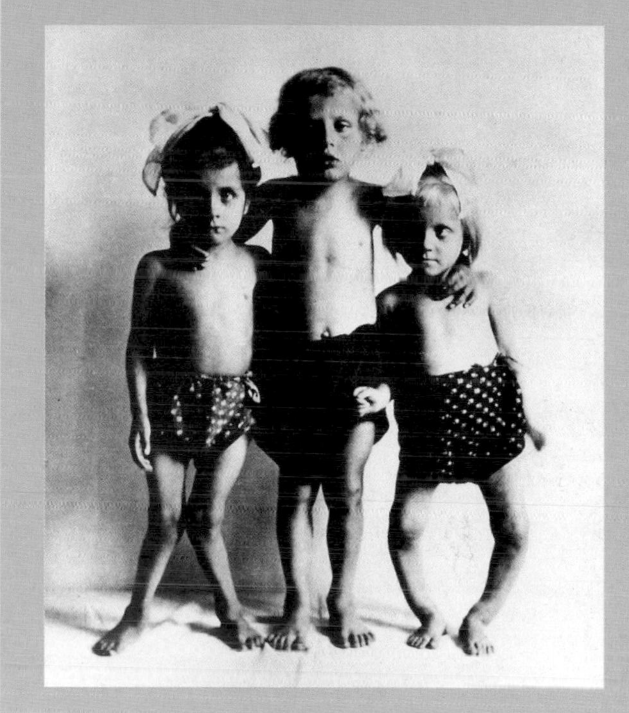

Teeth problems

Whatever food you eat, it sticks to your teeth. Sugary foods stick especially well. Bacteria in your mouth break down the sugar and make acid, which eats into the surface of your teeth and causes damage. So try not to eat too much sugar – and brush your teeth regularly.

Why not stay in bed all day?

I think we should be allowed to stay in bed all day if we want to. There are lots of reasons why this is a good idea.

It's fun to stay in bed. You can do most of the things you'd do out of bed – watch television, play computer games, read books or comics. You don't need to be up for any of these things.

There is no need to get dressed if you are in bed. So that saves on washing and on clothes.

Scientists say that people are taller when they get up in the morning than when they go to bed at night. This is because the cartilage between the bones in your spine stretches out when there is no pressure on it. So if you stay in bed all day, you must get even taller.

Lastly, everyone knows that the human body needs rest. Our parents chase us to bed every night, saying it is good for us. So why not do it all day too?

Therefore it is clear there are lots of reasons to stay in bed all day: lots to do there, savings on clothes and washing, letting our body rest and getting taller. Pass the duvet...

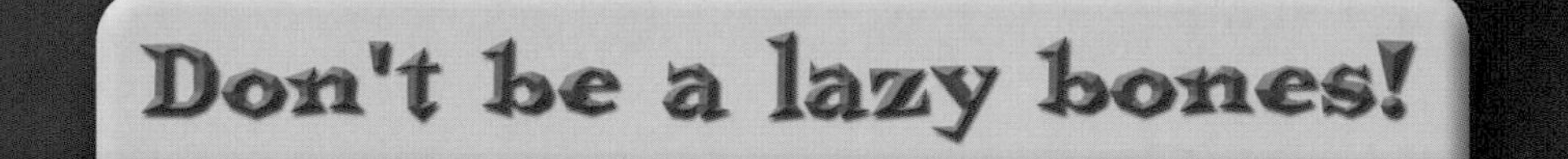

Too much rest is bad

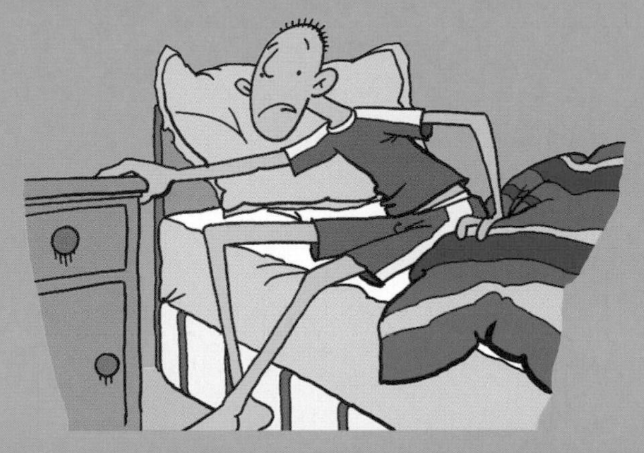

- not using muscles makes them weak
- exercise prevents this

Bones get thin if they don't bear weight

- astronauts' bones get thin in Space
- increased risk of breaks

Lack of exercise

- low energy
- boredom and depression

What bones can tell us

12 September 2002

DAILY NEWS

Features Section

SECRETS OF BONES:

Archaeology gets a Human Face

When most of us think about archaeology, we imagine people fitting together bits of old pots. However, today's archaeologists are just as likely to be piecing together a picture of our ancestors, based on the bones they find.

by Juliet Allen

Professor Williamson examines bones found in a recent dig

It's always an exciting moment in an archaeological dig when skeletons are found. The bones may be hundreds of years old but the way they have been buried can tell us a lot about our ancestors, and the customs of past civilizations.

By carefully examining the bones themselves, archaeologists can also find out all sorts of things about the lives of individual people.

A painstaking process

'We take the bones away to a laboratory, and clean them very thoroughly,' explains archaeology professor Jane Williamson. 'Then, through a mixture of careful observation and exact measurements, we slowly piece together a picture of the individuals whose remains we have found.'

Even though the bones may belong to people who died hundreds of years ago, the professionals can find out many details about them. These include their age, gender, height – and sometimes even why they died.

DAILY NEWS 12 September 2002

What bones can tell an expert

Age

Children: aged 1–2 years

- there are certain teeth that grow at particular times

Adults: young, middle-aged or old

- long bones, e.g. leg bones, stop growing at about age 24 – then the growth area changes appearance
- permanent teeth have all appeared by about age 30, then few changes
- if there are signs of arthritis this shows a person was old
- teeth wear down with age – but rate depends on what foods are eaten and this is not always known

Height

Database of measurements shows that length of thigh bone indicates the person's height

Life history/cause of death

- broken bones might show the cause of death
- mended bones show earlier breaks
- some diseases can scar bones (e.g. TB, cancer)

Gender

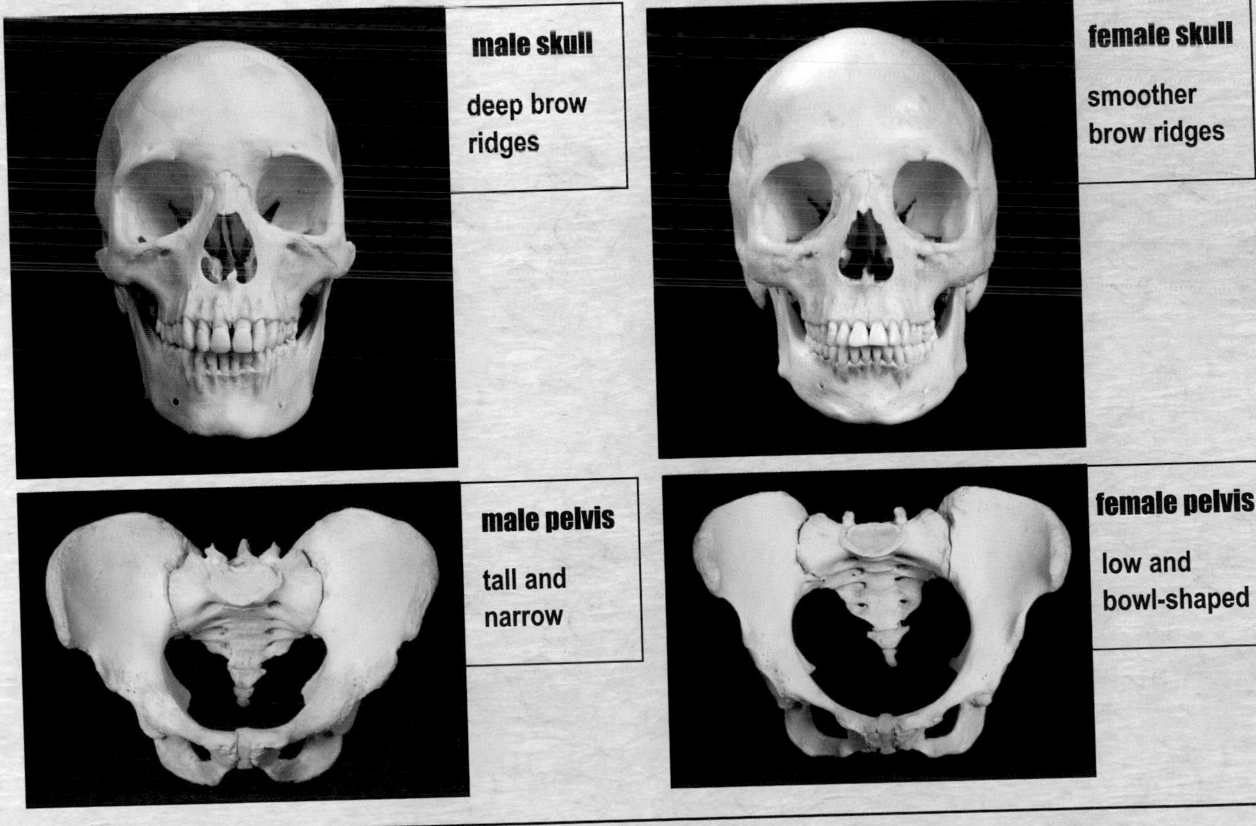

Mary Anning, dinosaur hunter

21 May 1799: born, Lyme Regis, Dorset

father: cabinet maker and fossil hunter

1810: father dies

Family left poor and owing money.

1811: with brother Joseph, discovers ichthyosaur fossil – first in Britain.

It is sold for £23 – enough to feed family for 6 months.

1817: very poor again

Lieutenant-Colonel Thomas Birch (wealthy fossil collector) sells his own fossil collection and gives money to Anning family.

Fossils

Creatures which lived millions of years ago are sometimes preserved. Their bones are slowly converted into rock, becoming what we call fossils. The rock versions of the bones show us the exact shape of the creature.

Fossil hunting became a craze in the 1800s and many fossils were found. Scientists studying the fossils began to suggest that mankind came into existence, not as was written in the Bible, but by a process of evolution slowly, over millions of years, while earlier creatures, like the dinosaurs, died out.

Ichthyosaur fossil

1832: ichthyosaur found by Mary and now in the Natural History Museum.

present day: many museums have Mary's fossils on display

Her name often not credited: because a woman or because she sold fossils rather than donated them?

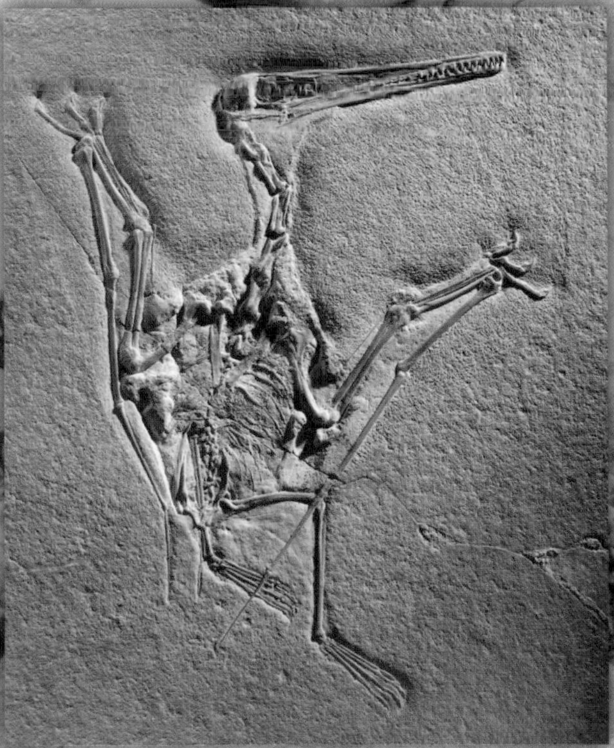

1820s: Mary takes charge of fossil business, selling fossils to museums, collectors, etc. She is celebrated by London scientists.

1823: finds a plesiosaur – sold for £125

9 March 1847: dies of breast cancer

1828: finds a pterodactyl

Pterodactyl fossil

Plesiosaur fossil

Running a marathon

Had to get fit first

1 year before

visited gym

ran short distances (1.5–8 km)

improved my diet: less fat, more protein and carbohydrate, more fruit and vegetables

started to drink more water: 1–2 litres a day

1 month before

ran 5–6 times a week

shortest runs: 8 km longest: 32 km

2 months before

ran 4–5 days a week

shortest runs: 8 km; longest: 24 km

6 months before

ran 4 days a week

distances increasing: 8–16 km

started to drink a lot more water: 2+ litres a day

Marathons are a very long race – over 42.2 km (26 miles). Only adults can run marathons because their heart, lungs, bones and muscles are fully developed. A younger body could not cope with the demands of such a long run.

1 week before

short runs only: less than 8 km

eating good diet

the day before

ate lots of carbohydrates for lunch, smaller dinner

went to bed early

the big day

8.00 a.m.
ate light breakfast and energy drink

10.00 a.m.
started running

12.00 noon
drank water and energy drinks

3.00 p.m.
the finish line

put on a silver blanket to stop me getting too cold

very tired – very happy!

Never again

next day

all muscles ached

very tired

Maybe in 2 years' time

day after that

Bones and muscles facts

Baby bones

Babies don't have bones at first. They have a sort of pattern, laid down in cartilage, which is slowly replaced by bone as the baby develops and grows.

Biggest and smallest

Biggest bone: the shoulder blade

Largest muscles: the gluteus muscles that are our buttocks

Longest bone: the femur (thigh bone)

Smallest bones: the hammer, anvil and stirrup, inside your ear

Smallest muscles: the muscles attached to the bones in the inner ear

Cartilage

Cartilage is a **flexible**, rubbery substance that protects the ends of bones and joints.

Some things we think are bone are made of cartilage, for example, our ears and the end of our nose.

Some creatures, such as sharks, have cartilage skeletons and no bones at all.

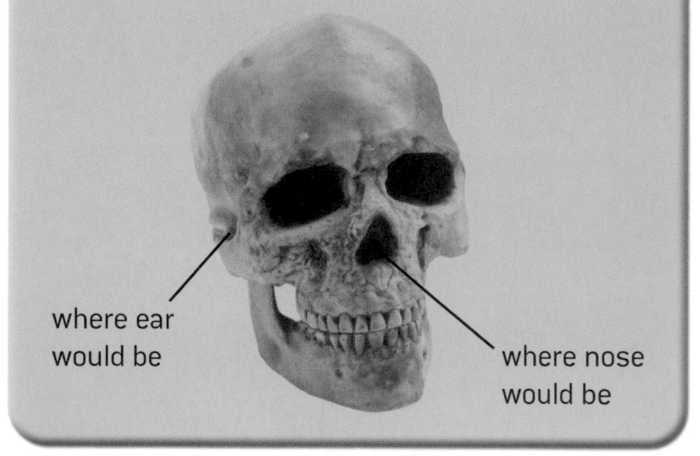

Latin names

All our bones have medical names. Here are some of them:

ankle	tarsals
backbone	vertebrae
collar bone	clavicle
fingers/toes	phalanges (say *fal-**an**-jees*)
hips	pelvis
knee cap	patella
shin bone	tibia
shoulder blade	scapula
skull	cranium (say ***crane**-ee-um*)
thigh bone	femur

cranium
clavicle
vertebrae
scapula
pelvis
phalanges
femur
patella
tibia
tarsals
phalanges

Muscles

The fibres that make up human muscles can be up to 30 cm long.

Numbers

Adults have about 206 bones.

Babies have about 300 bones. Some bones fuse together as they grow.

Giraffes have seven neck bones – the same number as humans! But they are a lot longer and more spaced out.

People have about 650 muscles.

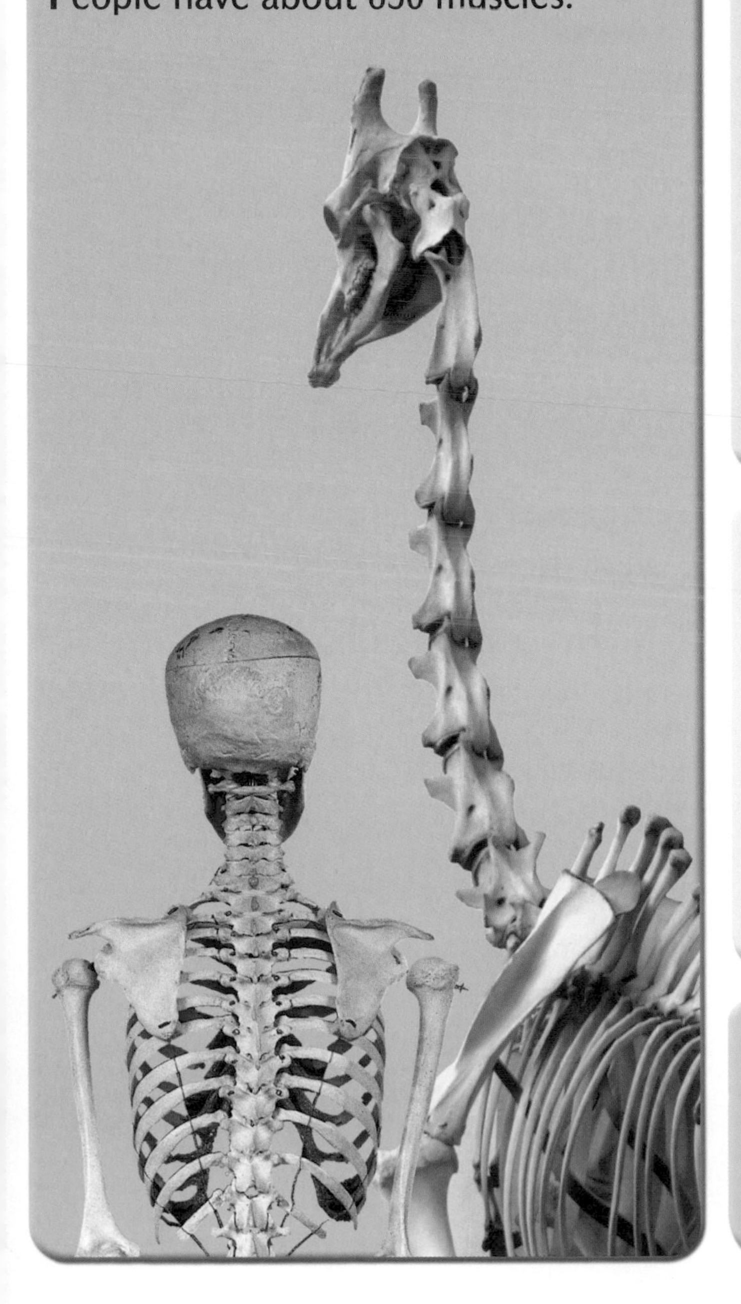

Relax and smile

Your face is covered in muscles. It takes 43 muscles to frown and 17 to smile. So smile! It's less work.

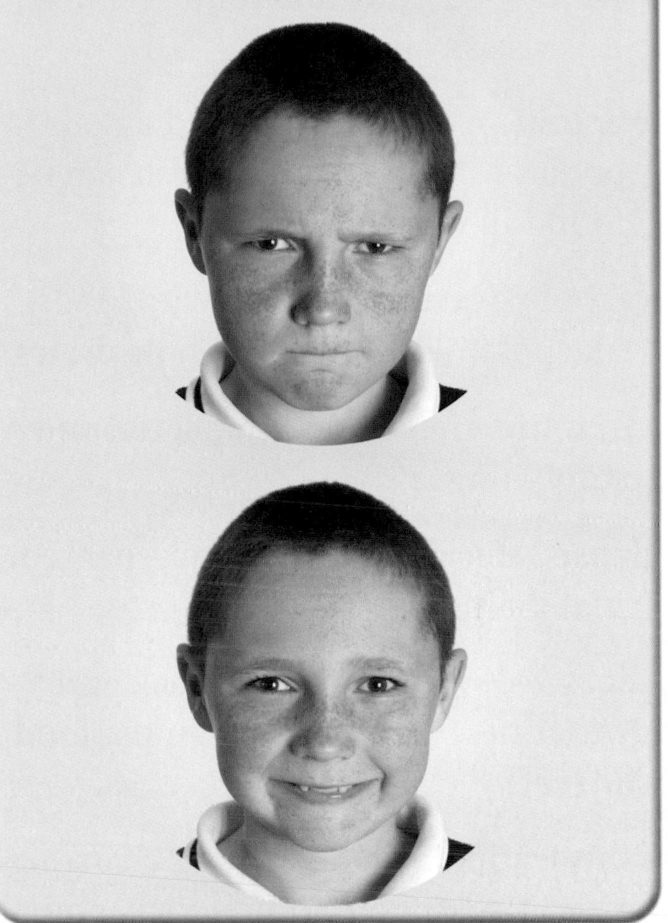

Strongest muscle

Probably the strongest muscle in the body is the muscle that makes up the heart. It can never rest, even when you are asleep, but it never gets tired. This muscle is so strong that it could squash a tennis ball every time it beats.

Weirdest bones

There are three tiny bones in your inner ear, which hit each other and help you to hear.

Glossary

arthropod an invertebrate with a segmented body and jointed limbs, e.g. insects and spiders

calcium a mineral required by our bodies for growth and which can be found in dairy products

carbohydrates energy giving foods, like potatoes, contain carbohydrates

circumference the distance around something

dense thick, made of closely-packed material

digestive system the stomach and intestines, that break down the food we eat

evaporation the process of changing from a solid or a liquid to a vapour or gas

exoskeleton a hard skeleton on the outside of a creature

external on the outside

flexible able to bend without breaking

fracture a break in a bone

internal inside

invertebrate a creature without a backbone

limbs legs, arms or wings

marrow a soft tissue in the middle of bones, which makes new blood cells

microscopic too small to see with the naked eye

organ a part of the body which has a particular job to do, like the brain or the heart

proportion the relation of one thing to another in size

rigid stiff, not possible to bend

sheath a close-fitting outer layer

shed to make or let something fall off

spongy with a loose structure full of holes, like a sponge

temporary designed to last for a short time only

tendon a strong fibre, like rope, which connects muscles to bones

vertebrate a creature with a backbone

vitamins these come in many different forms but are all very important for the health of our bodies

vulnerable open to attack or injury

Bibliography

Books

Non-fiction

Baldwin, D. *Health and Exercise (Your Health)*
ISBN: 08507 89974

Heddle, R. and Davies, K. *Science and your Body*
ISBN: 07460 14252

Johnson, J. *Under the microscope: Skeleton, our body's framework*
ISBN: 07496 44001

Parker, S. *Look at your body: Skeleton*
ISBN: 07496 41118

Taylor, K. *Discover Science: Structure*
ISBN: 1841 386219

Ward, B. *The Skeleton and Movement*
ISBN: 08631 37075

Fiction

Carroll, L. *Alice in Wonderland*
ISBN: 0140620869

White, E. B. *Charlotte's Webb*
ISBN: 0140301852

Internet

health websites

www.bbc.co.uk/health/kids/muscles/shtml

www.bbc.co.uk/health/kids/activity/shtml

archaeology websites

www.spoilheap.co.uk

websites with information on Mary Anning:

www.dinosaur.org/dinotimemachine.htm

www.lymeregis.com/history/geolog_fossil.htm

Index

Anning, Mary 40–41
archaeology 38–39
arm 4, 5, 16–17, 24–25, 26–27
arthropods 32–33, 46

babies 12, 44, 45
backbone 4, 5, 14, 44
ball and socket joint 9
bird 4, 6
blood vessels 7, 20
brain 4, 24–25
breaking bones 16–17, 18–19, 20, 31, 37, 39
breathing 23
butterfly 14, 15
biceps 24–25

calcium 34, 35, 46
carbohydrates 34, 46
cartilage 36, 44
contracting 24–25
crab 4, 14

dinosaurs 40–41

ear 44, 45
elbow 8
energy 15, 34–35, 37
exercise 25, 28–29, 30–31, 37
exoskeleton 14, 15, 32–33, 46
external skeleton 4, 32–33

face 45
finger 8, 23, 44
fitness 28–29, 30–31
fish 4, 5, 6
food 23, 34–35, 42–43
fossil 40–41
fracture 16, 18–19, 46
frown 45

grasshopper 14, 32
growth 7, 12–13, 14–15, 34–35

head 10–11, 12–13
health 28–29, 34–35
heart 4, 23, 42, 45
hinge joint 8
hip 9, 44

internal skeleton 4, 32–33
invertebrate 4, 14, 32–33, 46

joints 8–9, 44

knee 6, 8, 44

leg 4, 5, 12–13, 25, 39
lungs 4

marathon 42–43
marrow 7, 46
measuring 10–11, 12–13
mending bones 7, 16–17, 20–21, 33, 35, 39
muscles 22–23, 24–25, 26–27, 28, 31,34–35, 43, 44–45

neck 45
nose 44

organ 4, 32, 33, 46
overweight 28, 30, 34–35

pelvis 39
plaster cast 16, 17, 21

rest 36–37
ribs 4, 5
rickets 35

shoulder 9, 44
skeleton 4, 5, 8, 14, 32–33, 38–39
skin 14, 19, 22, 32–33
skull 4, 39, 44
slug 14, 22
smile 45
snake 5
spider 14, 32
spine 4, 36
sugar 35

teeth 35, 39
tendon 23, 46
thigh 9, 44
triceps 24–25

vertebrate 4, 14, 32–33, 46
vitamins 34–35, 46

worm 14, 22

x-ray 16